MATERIAL SCIENCE: BUILDING THE FUTURE

JOINT HEARING

BEFORE THE

SUBCOMMITTEE ON ENERGY &
SUBCOMMITTEE ON RESEARCH AND TECHNOLOGY

COMMITTEE ON SCIENCE, SPACE, AND TECHNOLOGY

HOUSE OF REPRESENTATIVES

ONE HUNDRED FIFTEENTH CONGRESS

FIRST SESSION

June 28, 2017

Serial No. 115–19

Printed for the use of the Committee on Science, Space, and Technology

Available via the World Wide Web: http://science.house.gov

U.S. GOVERNMENT PUBLISHING OFFICE

26–236PDF WASHINGTON : 2017

For sale by the Superintendent of Documents, U.S. Government Publishing Office
Internet: bookstore.gpo.gov Phone: toll free (866) 512–1800; DC area (202) 512–1800
Fax: (202) 512–2104 Mail: Stop IDCC, Washington, DC 20402–0001

COMMITTEE ON SCIENCE, SPACE, AND TECHNOLOGY

HON. LAMAR S. SMITH, Texas, *Chair*

FRANK D. LUCAS, Oklahoma	EDDIE BERNICE JOHNSON, Texas
DANA ROHRABACHER, California	ZOE LOFGREN, California
MO BROOKS, Alabama	DANIEL LIPINSKI, Illinois
RANDY HULTGREN, Illinois	SUZANNE BONAMICI, Oregon
BILL POSEY, Florida	ALAN GRAYSON, Florida
THOMAS MASSIE, Kentucky	AMI BERA, California
JIM BRIDENSTINE, Oklahoma	ELIZABETH H. ESTY, Connecticut
RANDY K. WEBER, Texas	MARC A. VEASEY, Texas
STEPHEN KNIGHT, California	DONALD S. BEYER, JR., Virginia
BRIAN BABIN, Texas BARBARA	JACKY ROSEN, Nevada
COMSTOCK, Virginia	JERRY MCNERNEY, California
GARY PALMER, Alabama	ED PERLMUTTER, Colorado
BARRY LOUDERMILK, Georgia	PAUL TONKO, New York
RALPH LEE ABRAHAM, Louisiana	BILL FOSTER, Illinois
DRAIN LaHOOD, Illinois	MARK TAKANO, California
DANIEL WEBSTER, Florida	COLLEEN HANABUSA, Hawaii
JIM BANKS, Indiana	CHARLIE CRIST, Florida
ANDY BIGGS, Arizona	
ROGER W. MARSHALL, Kansas	
NEAL P. DUNN, Florida	
CLAY HIGGINS, Louisiana	

SUBCOMMITTEE ON ENERGY

HON. RANDY K. WEBER, Texas, *Chair*

DANA ROHRABACHER, California	MARC A. VEASEY, Texas, *Ranking Member*
FRANK D. LUCAS, Oklahoma	ZOE LOFGREN, California
MO BROOKS, Alabama	DANIEL LIPINSKI, Illinois
RANDY HULTGREN, Illinois	JACKY ROSEN, Nevada
THOMAS MASSIE, Kentucky	JERRY MCNERNEY, California
JIM BRIDENSTINE, Oklahoma	PAUL TONKO, New York
STEPHEN KNIGHT, California, Vice Chair	JACKY ROSEN, Nevada
DRAIN LaHOOD, Illinois	BILL FOSTER, Illinois AMI
DANIEL WEBSTER, Florida	BERA, California MARK
NEAL P. DUNN, Florida	TAKANO, California
LAMAR S. SMITH, Texas	EDDIE BERNICE JOHNSON, Texas

SUBCOMMITTEE ON RESEARCH AND TECHNOLOGY

HON. BARBARA COMSTOCK, Virginia, *Chair*

FRANK D. LUCAS, Oklahoma	DANIEL LIPINSKI, Illinois
RANDY HULTGREN, Illinois	ELIZABETH H. ESTY, Connecticut
STEPHEN KNIGHT, California	JACKY ROSEN, Nevada
DARIN LaHOOD, Illinois	SUZANNE BONAMICI, Oregon
RALPH LEE ABRAHAM, Louisiana	AMI BERA, California
DANIEL WEBSTER, Florida	DONALD S. BEYER, JR., Virginia
JIM BANKS, Indiana	EDDIE BERNICE JOHNSON, Texas
ROGER W. MARSHALL, Kansas	
LAMAR S. SMITH, Texas	

CONTENTS

June 28, 2017

	Page
Witness List	2
Hearing Charter	3

Opening Statements

Statement by Representative Randy K. Weber, Chairman, Subcommittee on Energy, Committee on Science, Space, and Technology, U.S. House of Representatives	4
Written Statement	6
Statement by Representative Marc A. Veasey, Ranking Member, Subcommittee on Energy, Committee on Science, Space, and Technology, U.S. House of Representatives	8
Written Statement	10
Statement by Representative Barbara Comstock, Chairwoman, Subcommittee on Research and Technology, Committee on Science, Space, and Technology, U.S. House of Representatives	12
Written Statement	14
Statement by Representative Daniel Lipinski, Ranking Member, Subcommittee on Research and Technology, Committee on Science, Space, and Technology, U.S. House of Representatives	16
Written Statement	18

Witnesses:

Dr. Matthew Tirrell, Deputy Laboratory Director for Science and Chief Research Officer, Argonne National Laboratory	
Oral Statement	21
Written Statement	23
Dr. Laurie Locascio, Acting Associate Director for Laboratory Programs and Director, Material Measurement Laboratory, National Institute of Standards and Technology	
Oral Statement	29
Written Statement	41
Dr. Adam Schwartz, Director, Ames Laboratory	
Oral Statement	39
Written Statement	41
Dr. Fred Higgs, John and Ann Doerr Professor of Mechanical Engineering, Rice University	
Oral Statement	50
Written Statement	52
Discussion	62

Appendix I: Additional Material for the Record

Statement submitted by Representative Eddie Bernice Johnson, Ranking Member, Committee on Science, Space, and Technology, U.S. House of Representatives	84
Document submitted by Representative Bill Foster, Committee on Science, Space, and Technology, U.S. House of Representatives	86

MATERIAL SCIENCE: BUILDING THE FUTURE

WEDNESDAY, JUNE 28, 2017

House of Representatives,
Subcommittee on Energy and
Subcommittee on Research and Technology
Committee on Science, Space, and Technology,
Washington, D.C.

The Subcommittees met, pursuant to call, at 10:09 a.m., in Room 2318 of the Rayburn House Office Building, Hon. Randy Weber [Chairman of the Subcommittee on Energy] presiding.

LAMAR S. SMITH, Texas
CHAIRMAN

EDDIE BERNICE JOHNSON, Texas
RANKING MEMBER

Congress of the United States
House of Representatives
COMMITTEE ON SCIENCE, SPACE, AND TECHNOLOGY

2321 RAYBURN HOUSE OFFICE BUILDING
WASHINGTON, DC 20515–6301
(202) 225–6371
www.science.house.gov

Subcommittees on Energy and Research and Technology

Materials Science: Building the Future

Wednesday, June 28, 2017
10:00 a.m.
2318 Rayburn House Office Building

Witnesses

Dr. Matthew Tirrell, Deputy Laboratory Director for Science and Chief Research Officer, Argonne National Laboratory

Dr. Laurie Locascio, Acting Associate Director for Laboratory Programs and Director, Material Measurement Laboratory, National Institute of Standards and Technology

Dr. Adam Schwartz, Director, Ames Laboratory

Dr. Fred Higgs, John and Ann Doerr Professor of Mechanical Engineering, Rice University

U.S. HOUSE OF REPRESENTATIVES
COMMITTEE ON SCIENCE, SPACE, AND TECHNOLOGY

HEARING CHARTER

Wednesday, June 28, 2017

TO: Members, Committee on Science, Space, and Technology

FROM: Majority Staff, Committee on Science, Space, and Technology

SUBJECT: Energy Subcommittee and Research and Technology Subcommittee hearing: "Materials Science: Building the Future"

The Subcommittee on Energy and the Subcommittee on Research and Technology of the Committee on Science, Space, and Technology will hold a joint hearing titled *Materials Science: Building the Future* on Wednesday, June 28, 2017, at 10:00 a.m. in Room 2318 of the Rayburn House Office Building.

Hearing Purpose:

The purpose of the hearing is to review federally funded research in materials science, and highlight the broad economic impact of basic research in this area. New materials can improve the generation, storage, and use of energy, reduce the environmental impact and improve safety of energy production technologies, and provide the foundation for new technologies in medicine, transportation, manufacturing, protective systems, and computing. The hearing will also examine the benefits of user facilities, science prize competitions, and public-private partnerships to speed the development of advanced materials.

Witness List

- **Dr. Matthew Tirrell,** *Deputy Laboratory Director for Science and Chief Research Officer, Argonne National Laboratory*
- **Dr. Laurie Locascio,** *Acting Associate Director for Laboratory Programs and Director, Material Measurement Laboratory, National Institute of Standards and Technology*
- **Dr. Adam Schwartz,** *Director, Ames Laboratory*
- **Dr. Fred Higgs,** *John and Ann Doerr Professor of Mechanical Engineering, Rice University*

Staff Contact

For questions related to the hearing, please contact Emily Domenech of the Majority Staff at 202-226-2179.

Chairman WEBER. The Subcommittees on Energy and Research will come to order.

Without objection, the Chair is authorized to declare recesses of the Subcommittees at any time.

So welcome to today's hearing titled "Materials Science: Building the Future."

I now recognize myself for five minutes for an opening statement.

Today, we will have the opportunity to review federally funded research in materials science. I want to thank our panel of witnesses for joining us to share your important research, and provide the knowledge necessary to set priorities for basic science research.

Materials science is the discovery of new materials with novel structures, functions, and properties. In this area of science, researchers study the chemical, physical, atomic, and magnetic properties of an existing material, and use that knowledge to create new materials with ideal properties. By designing and creating new materials, researchers at our national labs and universities can solve complex engineering challenges and enable the development of new technologies.

Today, federal agencies ranging from the Department of Defense to the National Science Foundation and DOE are pursuing research in this area because the value to our end users is clear. By tailor-making materials for a specific use, scientists can create materials that increase efficiency and better store energy; reduce the environmental impacts and improve the safety of energy production technologies; develop stronger and more resilient artificial joints; improve high performance computing systems; and better protect our soldiers and athletes in the field.

As Madonna would say, we are certainly living in a material world. For example, Dr. Fred Higgs, who joins us from Rice University—my sister graduated from Rice, Dr. Higgs—and I were having that conversation—will testify about how the development of materials such as diamond-like carbons and nanocrystalline diamond can lead to long-lasting, wear-resistant artificial knees and hips that could last decades longer than today's technology.

At Ames Lab, led by Dr. Adam Schwartz who joins our panel today, the Department of Energy has cultivated decades of expertise in metallurgy and materials science. Researchers at Ames Lab pioneered the use of metallic powders in 3D printing. As Dr. Schwartz will testify, this expertise has enabled the production of high-purity metal powders that can be used in the creation of industrial parts for military, biomedical, and aerospace applications.

I'm also particularly interested in Ames' ongoing early-stage research in caloric materials for refrigeration and air conditioning— I own an air conditioning company, which if—and we're going to talk about this, in fact, the whole hearing may be on this— which if successful—I mean, how cool is that, right?—which if successful could save 20 to 25 percent of the generated electricity used for cooling, refrigeration, and air conditioning in the United States. Now, let that sink in: 20 to 25 percent of the energy used for refrigeration and air conditioning and heating in the United States.

Finally, just this week, a researcher at Argonne National Lab, which Dr. Tirrell is testifying on behalf of today, won the 2017 TechConnect National Innovation Award for developing a more efficient method to create graphene. This one area of materials science research could improve technology for advanced touch screens, long-lasting batteries, transparent and conducting coatings for solar cells, and next-generation oil-free solid lubricants.

Materials science also provides a perfect example of the broad economic benefit of investments in research infrastructure. The core capabilities and user facilities at our national labs are essential for the discovery and design of new materials. There is nowhere else in the world where an individual researcher or company could access a light source, high performance computing capabilities, and the specific expertise in materials synthesis that is available in our system of national labs.

You may hear today about how this vital area of research is at risk of being left behind because of budget cuts or changing priorities but basic and early stage research in materials science is exactly what this Committee has always supported.

Discoveries in materials science require tools and expertise provided by national labs, and industry users are ready and waiting to commercialize—commercialize—they're waiting to take it to market technology based on this fundamental science.

Hearings like today's help remind us of the Science Committee's core focus: the basic research that provides the foundation for technology breakthroughs. Before we can ever see the deployment of a better battery, a stronger material for protective gear, or wear-resistant materials for medicine or energy production, we must invest in the science infrastructure that makes these discoveries possible.

[The prepared statement of Chairman Weber follows:]

COMMITTEE ON
SCIENCE, SPACE, & TECHNOLOGY
Lamar Smith, Chairman

For Immediate Release
June 28, 2017

Media Contact: Kristina Baum
(202) 225-6371

Statement of Chairman Randy Weber (R-Texas)
Material Science: Building the Future

Chairman Weber: Today, we will have the opportunity to review federally funded research in materials science. I want to thank our panel of witnesses for joining us to share your important research, and provide the knowledge necessary to set priorities for basic science research.

Materials science is the discovery of new materials with novel structures, functions, and properties. In this area of science, researchers study the chemical, physical, atomic, and magnetic properties of an existing material, and use that knowledge to create new materials with ideal properties. By designing and creating new materials, researchers at our national labs and universities can solve complex engineering challenges and enable the development of new technologies.

Today, federal agencies ranging from the Department of Defense to the National Science Foundation and DOE are pursuing research in this area – because the value to the end user is clear. By tailor making materials for a specific use, scientists can create materials that increase efficiency and better store energy; reduce the environmental impacts and improve the safety of energy production technologies; develop stronger and more resilient artificial joints; improve high performance computing systems; and better protect our soldiers and athletes in the field. We are certainly living in a material world!

For example, Dr. Fred Higgs, who joins us from Rice University, will testify about how the development of materials such as diamond-like carbons and nanocrystalline diamond can lead to long-lasting, wear-resistant artificial knees and hips – that could last decades longer than today's technology.

At Ames Lab, led by Dr. Adam Schwartz who joins our panel today, the Department of Energy has cultivated decades of expertise in metallurgy and materials science. Researchers at Ames Lab pioneered the use of metallic powders in 3D printing. As Dr. Schwartz will testify, this expertise has enabled the production of high purity metal powders that can be used in the creation of industrial parts for military, biomedical, and aerospace applications.

I'm also particularly interested in Ames' ongoing early stage research in caloric materials for refrigeration and air conditioning – which if successful could save 20 to

25% of the generated electricity used for cooling, refrigeration, and air conditioning in the U.S.

Finally, just this week, a researcher at Argonne National Lab – which Dr. Tirrell is testifying on behalf of today – won the 2017 TechConnect National Innovation Award for developing a more efficient method to create graphene.

This one area of materials science research could improve technology for advanced touch screens, long-lasting batteries, transparent and conducting coatings for solar cells and next-generation oil-free solid lubricants.

Materials science also provides a perfect example of the broad economic benefit of investments in research infrastructure. The core capabilities and user facilities at our national labs are essential for the discovery and design of new materials. There is nowhere else in the world where an individual researcher or company could access a light source, high performance computing capabilities, and the specific expertise in materials synthesis that is available in our system of national labs.

You may hear today about how this vital area of research is at risk of being left behind because of budget cuts or changing priorities. But basic and early stage research in materials science is exactly what this Committee has always supported.

Discoveries in materials science require tools and expertise provided by national labs – and industry users are ready and waiting to commercialize technology based on this fundamental science.

Hearings like today's help remind us of the Science Committee's core focus – the basic research that provides the foundation for technology breakthroughs. Before we can ever see the deployment of a better battery, a stronger material for protective gear, or wear-resistant materials for medicine or energy production, we must invest in the science infrastructure that makes these discoveries possible.

###

Chairman WEBER. I now yield to the Ranking Member, Mr. Veasey.

Mr. VEASEY. Thank you, Mr. Chairman, my fellow Texan, and also Chair Comstock, for holding this hearing. We have a very impressive panel today, and I want to thank each and every one of you being here. I'm going to make my remarks brief because I think that everybody's really interested to hear what they have to say today, and I'm sure that as you are aware, we'd be hard-pressed to find a scientific field that doesn't rely on materials science at some level to accomplish its research objectives. It is critically—it is a critically important area of research for answering the most pressing scientific questions and advancing our economy in the 21st century. Lightweight vehicles, high-performance building materials, more efficient turbines, and solar panels are just a few examples. The research and development of new materials can provide a direct benefit to consumers with savings on energy bills and benefits to our environment.

Scientists at universities, national laboratories, and in the private industry utilize federal research grants and scientific user facilities to explore the frontiers of materials research. A better understanding of the properties of ceramics, glass, metals, composites, polymers, and plastics is achieved through materials research. By optimizing these properties, we can address key hurdles in developing new technologies with a variety of applications. Energy efficiency and reliability, public health and safety, and environmental stewardship can all benefit from strong investments in material research. In fact, I think we could sit here all day and talk about the immense benefits of material research, and I know that we're going to do just that, and like I said a little bit earlier, I think everybody is really excited to hear what you have to say.

And while there seems to be strong support for this work in Congress, we cannot have this conversation without acknowledging the shortsighted and harmful Trump budget released last month. The Administration's budget would absolutely decimate the all-important field of materials science in the United States. The budget would cut sustainable transportation and renewable energy by 70 percent and energy efficiency by 80 percent. It would cut critical research on the electric grid and fossil fuels in half. It would eliminate ARPA–E, cut the Office of Sciences by 17 percent, and nuclear energy by 30 percent. All of these programs help fund the materials research that we will hear about today. And even if we wanted to, we can't balance the budget by slashing our research funding.

The Administration's budget proposal will make the United States less competitive. These cuts would cause us to lose jobs, harm our public health, and hurt our international R&D partnerships. The proposed cuts are just absolutely puzzling. They just make no sense.

I look forward to hearing from each of you on how the proposed budget cuts at DOE, at NSF, at NIST could hurt us in the area of materials research enterprise and U.S. competitiveness. I am particularly interested in hearing from Dr. Schwartz about the consequences these severe cuts could have at his laboratory, which has a special focus on materials research.

The Administration has claimed that the private sector would simply start funding these key research areas once the federal government cuts them from its budget but I don't think that's based in reality. In fact, Administration officials recently confirmed that have not even begun a conversation with the private sector to determine what industry would be able or willing to pick up. So let's get back to reality and continue our strong support for these high-value research programs that are vital for American competitiveness, our quality of life, and our scientific leadership.

And before I conclude, I do want to apologize to the Chair and the other Members and our panelists that are here today. We have an Armed Services markup today downstairs and so I'm going to be back and forth, but again, I think that what we're going to hear today is really going to be good and interesting, and I really appreciate the panelists that are here today.

Mr. Chairman, I yield back.

[The prepared statement of Mr. Veasey follows:]

OPENING STATEMENT
Ranking Member Marc Veasey (D-TX)
of the Subcommittee on Energy

House Committee on Science, Space, and Technology
Subcommittee on Energy
Subcommittee on Research & Technology
"Materials Science: Building the Future"
June 28, 2017

Thank you, Chairman Weber and Chairwoman Comstock, for holding this hearing. We have assembled an impressive panel today. Thank you all for being here. I will make my remarks brief so we can hear each of your testimonies.

As I'm sure you're aware, we'd be hard-pressed to find a scientific field that doesn't rely on materials science at some level to accomplish its research objectives. It's a critically important area of research for answering the most pressing scientific questions and advancing our economy in the 21st century. Lightweight vehicles, high-performance building materials, more efficient turbines, and solar panels are just a few examples. The research and development of new materials can provide a direct benefit to consumers with savings on energy bills and benefits to our environment.

Scientists at universities, national laboratories, and in the private industry, utilize federal research grants and scientific user facilities to explore the frontiers of materials research. A better understanding of the properties of ceramics, glass, metals, composites, polymers, and plastics is achieved through materials research. By optimizing these properties, we can address key hurdles in developing new technologies with a variety of applications. Energy efficiency and reliability, public health and safety, and environmental stewardship can all benefit from strong investments in material research. In fact, I think we could sit here and talk about the immense benefits of materials research all day. I assume we will spend a good part of this hearing doing just that.

While there seems to be strong support for this work in Congress, we cannot have this conversation without acknowledging the shortsighted and harmful Trump budget released last month. The Administration's budget would absolutely decimate the all-important field of materials science in the United States. The budget would cut sustainable transportation and renewable energy by 70% and energy efficiency by 80%. It would cut critical research on the electric grid and fossil fuels in half. It would eliminate The Advanced Research Projects Agency-Energy (ARPA-E), cut the Office of Science by 17%, and nuclear energy by 30%. All of these programs help fund the materials research that we will hear about today.

The Administration's budget proposal will make the United States less competitive. These proposed cuts are puzzling. I look forward to hearing from each of you on how the proposed budget cuts at DOE, NSF, and NIST could impact the materials research enterprise and U.S. competitiveness. I am particularly interested in hearing from Dr. Schwartz about the consequences these severe cuts could have at his laboratory, which has a special focus on materials research.

The Administration has claimed that the private sector would simply start funding these key research areas once the federal government cuts them from its budget. But this is not based in reality. In fact, Administration officials recently confirmed that have not even begun a conversation with the private sector to determine what industry would be able or willing to pick up. Let's get back to reality and continue our strong support for these high-value research programs that are vital for American competitiveness, our quality of life, and our scientific leadership.

Thank you again for testifying.

And thank you, Mr. Chairman. I yield back.

Chairman WEBER. Well, thank you, Marc. I appreciate that. The good news is that the President doesn't have the last word. He may have the first tweet but not the last word. Did I say that out loud?

I now recognize the Chairwoman of the Subcommittee on Research and Technology, Mrs. Comstock, for her opening statement.

Mrs. COMSTOCK. Thank you. Good morning.

Today's hearing focuses on vital research in materials science. This basic and fundamental research provides the foundation for important new technologies in many fields including medicine, transportation, manufacturing, defense, energy, and computing, which ultimately helps improve our quality of life and grows the U.S. economy.

Behind every new innovation from the iPhone to the International Space Station is decades of work by engineers, physicists, and chemists, creating the new materials that make it possible.

Advances in materials science have been achieved in a variety of ways, from public-private partnerships, science prize competitions, and through investments made by the federal government, industry, and universities. By investing in STEM education and the research infrastructure necessary to advance this area of basic research, the federal government can fast-track the development of industry specific materials that benefit American consumers.

One recent example of a public-private partnership that I find of great interest is the NIST work alongside the National Football League, General Electric Company, and Under Armour to support an open innovation prize in search of advanced materials to better absorb or dissipate energy. The Head Health Challenge will lead to the improvement in performance of protective equipment, like helmets, to help and protect head safety for men and women in uniform; Americans who work in manufacturing, construction, and other industries; and those who participate in athletics, starting with children who participate in school sports. We have heard so much recently about the long-lasting impact of head injuries, how it might be connected to Alzheimer's and others. This is really exciting work that's going on.

This kind of partnership is particularly encouraging because we should be doing everything in our power to help protect the lives of those who put themselves on the line for our freedom and safety as well as American workers and, of course, our children in those ever-present sports that we know are wonderful for them but we want them to perform in them safely.

By investing in materials science research, we invest in both innovation and the livelihood of our citizens.

Manufacturing is another critical sector where material science innovation can help create efficiency in production. While scientists develop new materials in our national labs and universities, industry applies these new materials to improve manufacturing, and create new products that keep the United States competitive in the global economy.

As Chair of the Research and Technology Subcommittee, I am interested in learning more about NIST's work with manufacturers and other private industry partners on new materials testing and standards, as well as the National Science Foundation's investment in basic research at institutions like Rice University.

Taxpayer investment in basic and fundamental research, which the private sector can then develop and commercialize, provides significant rewards that improve our society and the lives of our citizens. We must ensure that this research ecosystem is a vibrant, functioning partnership to spur innovation and create new industries and, of course, more jobs.

Thank you to our expert witnesses for being here today, and I look forward to hearing your informative testimony.

[The prepared statement of Mrs. Comstock follows:]

For Immediate Release
June 28, 2017

Media Contact: Kristina Baum
(202) 225-6371

Statement of Chairwoman Barbara Comstock (R-Va.)
Material Science: Building the Future

Chairwoman Comstock: Today's hearing focuses on vital research in materials science. This basic and fundamental research provides the foundation for important new technologies in many fields including medicine, transportation, manufacturing, defense, energy, and computing, which ultimately helps improve our quality of life and grows the U.S. economy.

Behind every new innovation – from the I-Phone to the International Space Station – is decades of work by engineers, physicists, and chemists, creating the new materials that make it possible.

Advances in materials science have been achieved in a variety of ways, from public-private partnerships, science prize competitions, and through investments made by the federal government, industry, and universities. By investing in STEM education and the research infrastructure necessary to advance this area of basic research, the federal government can fast-track the development of industry specific materials that benefit American consumers.

One recent example of a public-private partnership that I find of great interest is the National Institute of Standards and Technology's (NIST) work alongside the National Football League, General Electric Company, and Under Armour to support an open innovation prize in search of advanced materials to better absorb or dissipate energy.

The Head Health Challenge will lead to the improvement in performance of protective equipment, like helmets, to help and protect head safety for men and women in uniform; Americans who work in manufacturing, construction, and other industries; and those who participate in athletics, starting with children who participate in school sports.

This kind of partnership is particularly encouraging because we should be doing everything in our power to help protect the lives of those who put themselves on the line for our freedom and safety as well as American workers and our children. By investing in materials science research, we invest in both innovation and the livelihood of our citizens, so it is crucial that the United States leads in these efforts.

Manufacturing is another critical sector where material science innovation can help create efficiency in production. While scientists develop new materials in our national labs and universities, industry applies these new materials to improve manufacturing, and create new products that keep the United States competitive in the global economy.

As Chair of the Research and Technology Subcommittee, I am interested in learning more about NIST's work with manufacturers and other private industry partners on new materials testing and standards, as well as the National Science Foundation's investment in basic research at institutions like Rice University.

Tax-payer investment in basic and fundamental research, which the private sector can then develop and commercialize, provides significant rewards that improve our society and the lives of Americans. We must ensure that this research ecosystem is a vibrant, functioning partnership to spur innovation and create new industries and more jobs.

Thank you to our expert witnesses for being here today, and I look forward to hearing your informative testimony.

###

Chairman WEBER. Thank you, Mrs. Comstock, and I recognize the Ranking Member of the Subcommittee on Research and Technology, Mr. Daniel Lipinski, for his opening statement.

Mr. LIPINSKI. Thank you, Chairman Weber and Chairwoman Comstock, for holding this hearing on federal investments in materials science research and the economic importance of these programs.

Materials science and engineering R&D is carried out across several federal agencies. This research, as we will hear more about this morning, has applications across many sectors, including energy, defense, transportation, and even human welfare, as Chairwoman Comstock mentioned, the better helmets that can be made to prevent traumatic brain injury.

Unfortunately, as the Office of Science and Technology Policy detailed in a 2011 paper, the time it takes to move a newly discovered advanced material from the lab to the marketplace remains much too long. That white paper was the genesis of the multi-agency Materials Genome Initiative, or MGI. The MGI is a public-private R&D partnership that seeks to accelerate the lab-to-market timeline through advances in computational techniques, more effective use of standards, and enhanced data management.

The Research and Technology Subcommittee, on which I serve as Ranking Member, focuses on NSF and NIST, so I want to spend a moment talking about the important materials research programs at those agencies. NSF participates in the MGI primarily through the Designing Materials to Revolutionize and Engineer our Future program. This program is building the fundamental knowledge base needed to increase the precision of new materials development, enabling a shift from trial and error to designing and producing materials with specific desired properties. NSF also contributes to MGI through the Cyber-Enabled Materials, Manufacturing, and Smart Systems Initiative. As part of this initiative, NSF launched the Materials Innovation Platforms program to develop transformative techniques and instrumentation that will improve understanding and discovery of new, complex material systems.

NIST scientists conduct research in all aspects of materials science, with the goal of developing better and new measurement and characterization tools and standards for advanced materials. The agency's major efforts on material science research are supported by the Material Measurement Laboratory, the national reference laboratory for measurements in the chemical, biological, and material sciences. In addition to its internal research program, NIST also established the Advanced Materials Center of Excellence at Northwestern University, Argonne National Laboratory, and the University of Chicago, to facilitate the collaboration with leading research institutes and industry. The Center supports the goals of the Materials Genome Initiative by developing computational tools and databases to support materials discovery and production. Finally, NIST manages the interagency Manufacturing USA initiative, which includes several institutes focused on advanced materials. I look forward to learning more about all of this work from Dr. Locascio.

I want to echo the comments of my fellow Ranking Member, Mr. Veasey, by expressing my concern about the Trump Administration's proposed budget cuts to materials R&D across the science agencies. Not only would these cuts cause us to lose out on the economic opportunities our materials research programs create. They would also do great harm to our nation's ability to stay at the cutting edge of materials science and the related health, energy storage, technology, and national security benefits that will be discussed today.

We have an excellent panel before us that can help us understand not only materials science itself, but also why our investments in this field are so important for the nation. The proposed 11 percent cut at NSF, the 13 percent cut to the labs at NIST, and the even more draconian cuts at DOE must not be enacted. Today's hearing will give us a few more reasons why we must reject the President's budget request if our nation is to stay scientifically and economically competitive, and I certainly appreciate Chairman Weber's comments about that budget and what Congress will do. Hopefully we will see robust funding for these programs.

So I look forward to the testimony and discussion this morning, and I thank the panelists for being here to share their expertise with us.

With that, I yield back.

[The prepared statement of Mr. Lipinski follows:]

OPENING STATEMENT
Ranking Member Dan Lipinski (D-IL)
of the Subcommittee on Research and Technology

House Committee on Science, Space, and Technology
Subcommittee on Energy
Subcommittee on Research and Technology
"Materials Science: Building the Future"
June 28, 2017

Thank you Chairman Weber and Chairwoman Comstock for holding this hearing on federal investments in materials science research and the economic importance of these programs.

Materials science and engineering R&D is carried out across several federal agencies. This research, as we will hear more about this morning, has applications across many sectors, including energy, defense, transportation, and even human welfare – for example, better helmets to prevent traumatic brain injury.

Unfortunately, as the Office of Science and Technology Policy detailed in a 2011 white paper, the time it takes to move a newly discovered advanced material from the lab to the marketplace remains much too long. That white paper was the genesis of the multi-agency Materials Genome Initiative, or MGI. The MGI is a public-private R&D partnership that seeks to accelerate the lab to market timeline through advances in computational techniques, more effective use of standards, and enhanced data management.

The Research and Technology Subcommittee, on which I serve as Ranking Member, focuses on NSF and NIST, so I want to spend a moment talking about the important materials research programs at those agencies. NSF participates in the MGI primarily through the Designing Materials to Revolutionize and Engineer our Future program. This program is building the fundamental knowledge base needed to increase the precision of new materials development, enabling a shift from trial and error to designing and producing materials with specific desired properties. NSF also contributes to MGI through the Cyber-Enabled Materials, Manufacturing, and Smart Systems Initiative. As part of this initiative, NSF launched the Materials Innovation Platforms program to develop transformative techniques and instrumentation that will improve understanding and discovery of new, complex material systems.

NIST scientists conduct research in all aspects of materials science, with the goal of developing better and new measurement and characterization tools and standards for advanced materials. The agency's major efforts on material science research are supported by the Material Measurement Laboratory, the national reference laboratory for measurements in the chemical, biological, and material sciences. In addition to its internal research program, NIST also established the Advanced Materials Center of Excellence at Northwestern University, Argonne National Laboratory, and the University of Chicago, to facilitate collaboration with leading research institutes and industry. The Center supports the goals of the Materials Genome Initiative by developing computational tools and databases to support materials discovery and production. Finally, NIST manages the interagency Manufacturing USA initiative, which includes several institutes focused on advanced materials. I look forward to learning more about all of this work from Dr. Locascio.

I want to echo the comments of my fellow Ranking Member, Mr. Veasey, by expressing my concern about the Trump Administration's proposed budget cuts to materials R&D across the science agencies. Not only would these cuts cause us to lose out on the economic opportunities our materials research programs create, they would also do great harm to our nation's ability to stay at the cutting edge of materials science and the related health, energy storage, technology, and national security benefits that will be discussed today. We have an excellent panel before us that can help us understand not only materials science itself, but also why our investments in this field are so important for the nation. The proposed 11% cut at NSF, the 13% cut to the labs at NIST, and the even more draconian cuts at DOE must not be enacted. Today's hearing will give us a few more reasons why we must reject the President's budget request if our nation is to stay scientifically and economically competitive.

I look forward to the testimony and discussion this morning, and I thank the panelists for being here to share their expertise with us. I yield back.

Chairman WEBER. I thank the gentleman.

It is now time for witness introductions, and I'm going to yield right back to Mr. Lipinski to introduce our first witness today.

Mr. LIPINSKI. Thank you, Mr. Chairman.

Dr. Matthew Tirrell is Deputy Laboratory Director for Science and Chief Research Officer at Argonne National Laboratory in my district. At Argonne, he is responsible for integrating the laboratory's research and development efforts in science and technology capabilities. He is also the Founding Director of the Institute for Molecular Engineering at the University of Chicago, which has a mission to translate advances in basic physics, chemistry, biology, and computation into tools to address important societal problems. The Institute recently partnered with Argonne and Fermi National Labs to create the Chicago Quantum Exchange, which aims to serve as an intellectual hub for the science and engineering of quantum information and to commercialize discoveries through the Polsky Center for Entrepreneurship at the University of Chicago.

Dr. Tirrell received his achelor's degree from Northwestern University, just as I did, in engineering, and his Ph.D. from University of Massachusetts-Amherst. His distinguished career has included faculty positions at the University of Minnesota, the University of California-Santa Barbara, University of California-Berkeley, and induction into the National Academy of Engineering and the American Academy of Arts and Sciences.

Welcome, Dr. Tirrell. We're happy to have him here today.

Chairman WEBER. Thank you, Mr. Lipinski.

Our second witness today is Dr. Laurie Locascio—is that right? Okay—Acting Associate Director for Laboratory Programs and Director for the Material Measurement Laboratory at the National Institute of Standards and Technology. Previously, Dr. Locascio served as Chief of the Biochemical Division in the Material Measurement Laboratory. She received a Bachelor's of Science degree in chemistry from James Madison University, a master's of science degree in bioengineering from the University of Utah, and a Ph.D. in toxicology from the University of Maryland at Baltimore. Welcome.

Our next witness is Dr. Adam Schwartz, Director at Ames Laboratory. He is also a Professor of Materials Science and Engineering in the College of Engineering at Iowa State University. Dr. Schwartz had over 20 years of materials science research and management experience at Lawrence Livermore National Laboratory prior to joining Ames Laboratory. He received a bachelor's degree and master's degree in metallurgical engineering as well as a Ph.D. in materials science and engineering from the University of Pittsburgh. Welcome.

Our last witness is Dr. Fred Higgs, a John and Ann Doerr Professional of Mechanical Engineering at Rice University, where my sister graduated from. Previously, he was a postdoctoral research fellow at Georgia Institute of Technology. Dr. Higgs received a B.S. in mechanical engineering, an M.S. in mechanical engineering, and a Ph.D. in mechanical engineering—you have a thing for mechanical engineering—from—pronounce that.

Dr. HIGGS. Rensselaer.

Chairman WEBER. Rensselaer Polytech Institute in Troy, New York, but you finally made it to Texas. So I told him he's a native Texan imported from Florida. So welcome. We're glad you here.

And Dr. Tirrell, I now recognize you for five minutes to present your testimony, and welcome to you as well.

TESTIMONY OF DR. MATTHEW TIRRELL, DEPUTY LABORATORY DIRECTOR FOR SCIENCE AND CHIEF RESEARCH OFFICER, ARGONNE NATIONAL LABORATORY

Dr. TIRRELL. Thank you. Chairman Weber, Chairwoman Comstock, Ranking Member Veasey, and Ranking Member Lipinski and Members of the Subcommittees, thank you for the opportunity to appear today to discuss the future of materials science from the perspective of the U.S. Department of Energy National Laboratories.

Argonne National Lab was founded as a chemistry, materials, and nuclear engineering lab in 1946 as the successor to the Manhattan Project's metallurgical lab at the University of Chicago. My colleagues at Argonne and across the national laboratories seem to improve the way this nation generates, distributes and uses energy. Materials science and engineering are essential to this pursuit and to many other sectors of importance to society. Bringing fundamental advances in material sciences to reality for the ultimate benefit of society requires investments at various stages of development.

Though the time scale is accelerating via powerful new predictive computational methods, many developed at DOE laboratories, there remains a long lead time from conception, discovery and synthesis of new materials to their ultimate useful application. Indeed, important discoveries in materials science arise often without any application in mind. National laboratories differ from universities in performing both basic and applied research in an environment where unmatched characterization facilities and capabilities for scale-up exist.

The process of taking a fundamental discovery or invention to the point that industry will invest in commercial development is a very non-linear one involving iteration between fundamental and applied research. Pushing basic science toward practical applications frequently raises new basic science questions that have to be addressed before useful results emerge.

The history of electrochemical research at Argonne leading to new materials and devices for energy storage is a case in point. Electrochemical energy storage and research—storage research and development spans the battery field from basic materials research all the way to prototyping.

The prototyping often reveals the need for new insight at the fundamental level and inspires new basic research. A specific example is the Energy Innovation Hub at Argonne, the Joint Center for Enter Storage Research, or JCESR. Founded in 2012, JCESR has united government, academic and industrial researchers from many disciplines in a major research project that combines discovery science, battery design, prototyping, and manufacturing science in a single highly interactive organization. JCESR as an example of collaborative basic research leading to proof of concept prototypes is one we aim to model in other materials research areas.

A second powerful example is in the area of quantum computing. The exponential expansion and the power of information technology, which we call Moore's Law, has catalyzed U.S. productivity and growth for over the last 50 years but, like much of our nation's aging infrastructure, this is now ending as roadmaps that have worked since the 1960s are now reaching their limits. The research and industrial communities are mobilizing to search for fundamentally new approaches to information processing. Quantum computing is based on exploiting subtle aspects of quantum physics for unprecedented new information technologies. These technologies implemented via materials design and development can handle computationally complex problems, provide communications security, sensing technologies in ways that are impossible with conventional hardware.

Recognizing this promise, other nations such as China, Canada and several European countries are investing heavily in quantum material science. Argonne in collaboration with the University of Chicago and Fermilab, and I might add, Ames Lab and NIST, are poised to compete and lead in this area.

Water research is a third example where basic materials science is needed. Water and energy are deeply interrelated. Cooling in power plants, hydraulic fracturing, petroleum refining, biofuel production account for the majority of water withdrawals and, conversely, water treatment and distribution represents large consumers of electricity. This water-energy interdependence is leading materials scientists to work on devising new membranes, sorbents, sensors, catalysts and surface treatments to enable step change in improvements in energy-water systems.

Across the lab complex, the commitment to materials science breakthrough means using every specialized tool at hand. At Argonne, we leverage the high-energy x-rays of the advanced photon source to see materials at the atomic level and the computing power of the Leadership Computing Facility for Materials Characterization and Simulation. Upgrades underway at each of these facilities will serve to increase their power.

So in summary, DOE labs are an enormous asset in pursuing the broad spectrum of materials science and engineering research.

Thank you for your time and attention to this topic, and of course will answer any questions you may have.

[The prepared statement of Dr. Tirrell follows:]

Testimony of Dr. Matthew Tirrell
Deputy Laboratory Director for Science, Argonne National Laboratory
before the
Subcommittee on Energy and the Subcommittee on Research and Technology
of the
U.S. House Committee on Science, Space, & Technology
June 28, 2017

Chairman Weber and Chairwoman Comstock, Ranking Member Veasey and Ranking Member Lipinski, and members of the subcommittees, thank you for the opportunity to appear before you today to discuss the future of materials science, as seen from the perspective of U.S. Department of Energy National Laboratories. I am Matthew Tirrell, deputy laboratory director and chief research officer of the Argonne National Laboratory located in Lemont, Illinois. I am also professor and founding dean of the Institute for Molecular Engineering at the University of Chicago. Prior to joining Argonne and the University of Chicago in 2011, I was a faculty member at the University of California, Berkeley, with an appointment as a faculty scientist at the Lawrence Berkeley National Laboratory's Materials Science Division. From 1999-2009, I was dean of the College of Engineering at the University of California, Santa Barbara and a member of the Materials Department faculty there, which was ranked as the top program in the country in the last National Academy of Sciences National Research Council report. I am currently co-chair of an on-going National Academy of Sciences National Research Council study of the Future of Materials Science, supported by the Department of Energy and the National Science Foundation.

Argonne National Laboratory was founded as a chemistry, materials and nuclear engineering laboratory in 1946, as the successor to the Manhattan Project's Metallurgical Laboratory. Since then, as part of the Department of Energy (DOE) network of national laboratories, Argonne has built on its original strengths and expanded its mission in response to national needs. My colleagues at Argonne, and across the DOE, its Office of Science, and the other national laboratories seek to improve the way this nation generates, distributes, and uses energy. Materials science and engineering are essential to this pursuit and to many other sectors of importance to society, including next generation information technologies, transportation and health care. The next chapter in this country's history—realization of greater mobility, prosperity and wellbeing—depends on fundamental, atomic level breakthroughs in materials. In working together to achieve those breakthroughs, the labs are using not only decades of materials and chemical science expertise but also bringing to bear the unparalleled power of their supercomputing and imaging resources. An effective national materials science and engineering program requires healthy, balanced and interactive efforts spanning basic science and technology, all materials classes and the four fundamental elements of the field: properties, performance, structure and composition, and synthesis and processing.

Bringing fundamental advances in materials science to reality for the ultimate benefit of society requires a continuum of investments at various stages of development. Though the timescale is accelerating via powerful new predictive computational methods, many developed at DOE laboratories, there remains a long lead-time from conception, discovery and synthesis of new materials to their ultimate useful application. National laboratories play a unique role in connecting basic research to eventual commercial technologies. They differ from universities in performing both basic and applied research in an environment where unmatched characterization facilities and capabilities for scale up exist. The process of taking a fundamental discovery or invention to the point that industry will invest in commercial development is a very nonlinear one involving cyclic iteration between fundamental and applied research. Development of basic science toward practical applications frequently raises new basic science questions that must be addressed before progress can be made. Indeed, it can be difficult sometimes to definitively categorize research efforts as either basic or applied. For example, a new field of manufacturing science is emerging in which new fundamental questions related to speed, dynamics and scale of manufacturing processes must be addressed.

The history of electrochemical research at Argonne leading to new materials and devices for energy storage is a case in point. Electrochemical energy storage research and development programs span the battery technology field from basic materials research and diagnostics to prototyping and post-test analyses. Building prototypes reveals the need for new insight at the fundamental level and inspires new basic research. Argonne's multidisciplinary team of world-leading researchers is working to develop a fundamental knowledge base for advanced energy storage technologies to aid the growth of the U.S. battery manufacturing industry, transition the U.S. automotive fleet to plug-in hybrid and electric vehicles, and enable greater and more flexible use of any energy source. A specific example is the Energy Innovation Hub located at Argonne, the Joint Center for Energy Storage Research. Founded in 2012, JCESR has united government, academic, and industrial researchers from many disciplines in a major research project that combines discovery science, battery design, research prototyping, and manufacturing collaboration in a single highly interactive organization. The JCESR example of collaborative basic science leading ultimately to proof-of-concept prototypes is one we hope to model in other materials science efforts.

We are building upon Argonne's historical leadership in electrochemistry to create a broad research, development and demonstration program centered on advanced energy storage materials and systems for both mobile and stationary applications. We develop more robust, cost-effective and higher-energy density lithium-ion and beyond lithium-ion battery technologies, using our science and engineering capabilities to develop storage materials that dramatically increase energy and power densities. Materials science is at the heart of improving the way we interact with energy as 21st century citizens. Breakthroughs will enable the type of reliable, high volume energy storage we need to make our electric grid more stable and give hybrid and all-electric vehicles longer range and greater safety. The related field of fuel cell research, based on new fundamental research in catalysis, is leading the way toward mobile and distributed energy generation based on hydrogen.

A second powerful example is in the area of quantum information science and technology. The exponential in the power of information technology-Moore's Law-has catalyzed US efficiency and growth over the last 50 years. However, like much of our nation's aging infrastructure, digital-powered productivity needs an upgrade as scientific breakthroughs from the 1950s and 1960s reach their technological limits. This jeopardizes the safety and security of the American people and threatens what has been the backbone of US economic growth over the past several decades.

As Moore's Law reaches its apex, the research and industrial communities are mobilizing to search for fundamentally new approaches to information processing. Quantum technologies, based on fundamental particles of nature such as individual atoms and photons, are natural targets for innovation, as they hold great promise to become the computers, networks and sensors of tomorrow. Quantum information science is based on exploiting subtle aspects of quantum physics to create valuable technology able to solve scientific challenges that current technology has been inadequate to surmount. These technologies, implemented via new materials design and development, can handle computationally complex problems, provide communication security, enhanced navigation, imaging and other sensing technologies in ways that are impossible with conventional hardware. Recognizing this promise, other nations, such as China, are investing heavily in quantum materials science. Argonne, in collaboration with the University of Chicago and Fermilab, as well as Ames Lab, are poised to compete and lead in this area with the necessary investment. As in the energy storage materials area, the interaction of applications with basic science will guide the science toward the optimum materials.

Water-related research is a third example where materials science comes to the fore. Water is a unique, indispensable resource essential not only for life, but also for practically all forms of economic activity. The food we eat, the clothes we wear, the goods we use, and even the electricity we consume all require water to make. Energy and water are deeply interrelated. Cooling in power plants, hydraulic fracturing, petroleum refining, and biofuel production account for a major fraction of water withdrawals; conversely, water treatment, distribution, and use represent the largest consumers of electricity. The pressure on our water resources, together with the techno-economic implications of energy-water interdependence, has highlighted a need for new materials with interfaces whose surface affinities, reactivity, and microstructures take advantage of novel properties of complex, aqueous environments. The materials science community is devising effective new membranes, sorbents, sensors, catalysts, surface treatments, and coatings with tailored functionality, based on fundamental predictive design of interfacial properties and mechanistic understanding of their interactions with complex and confined environments. The co-design of new materials and fluids to exploit specific material-fluid interactions will enable step-change improvements in the design and selectivity for purification, transformation, and transport processes in energy-water systems.

In addition to these examples, materials science and engineering is also responsible for new solar panel designs, high-performance sponges for oil absorption, nanofiber magnets, high-performance lubricants, and improved nuclear energy fuels and materials. In nuclear the focus is on verifying the safety of current light-water reactors and developing new, high-performance materials that promise to improve the economics and further enhance safety of advanced reactors. At Argonne, this research leverages our capability to design and develop materials for extreme conditions, as well as our nuclear engineering capability dating to the advent of our lab.

To advance in the next stage of materials science, DOE and its labs are fine-tuning their approach to basic research, redoubling their efforts to work together and digging deeper with the specialized scientific tools that are the hallmark of the lab complex. Across the lab complex, the commitment to materials science breakthroughs means using every specialized tool at hand. At Argonne, we leverage the high-energy X-rays of our Advanced Photon Source to see materials at the atomic level and the computing power of our Leadership Computing Facility for materials characterization and simulation—upgrades underway at both of these facilities will serve to increase their power. At our Center for Nanoscale Materials, we manipulate material interactions at the nanoscale and synthesizing nano-architectures for energy, information, and functionality. University collaborations are important. A number of projects funded by the DOE Office of Science, Basic Energy Sciences program and Office of Energy Efficiency & Renewable Energy, Vehicle Technologies Office, advance materials science across the research and development spectrum. The Midwest Integrated Center for Computational Materials, for example, develops and disseminates computational tools to simulate and predict properties of materials for energy conversion processes, while our Materials Engineering Research Facility uses cutting-edge tools to scale up production of newly discovered materials.

Thank you for your time and attention to this critically important topic. I would be pleased to respond to any questions that you might have.

Matthew Tirrell, PhD
Deputy Laboratory Director for Science and Chief Research Officer
Argonne National Laboratory

Matthew Tirrell is Deputy Laboratory Director for Science and Chief Research Officer at Argonne National Laboratory. He is responsible for integrating the laboratory's research and development efforts and science and technology capabilities. He develops and drives strategy to support integrated teams across disciplines in support of Argonne's strategic initiatives. He assists in reviews of scientific programs and chairs the safety Corrective Action Review Board.

Dr. Tirrell also serves as Founding Pritzker Director of the Institute for Molecular Engineering (IME) at the University of Chicago. Before becoming founding director of the IME in 2011, Dr. Tirrell was dean of engineering at the University of California in Santa Barbara. Previously he served as the Arnold and Barbara Silverman Professor and chair of the Department of Bioengineering at the University of California at Berkeley and as a professor of materials science and engineering and chemical engineering and faculty scientist at Lawrence Berkeley National Laboratory.

Dr. Tirrell began his academic career at the University of Minnesota, where he was the Shell Distinguished Chair in Chemical Engineering, Earl E. Bakken Professor of Biomedical Engineering, director of the Biomedical Engineering Institute, and head of Chemical Engineering and Materials Science.

He has provided leadership to numerous national and regional organizations, including the Science and Technology Panel of the University of California President's Council for National Laboratory Administration from 2000 to 2010.

Dr. Tirrell received his bachelor's degree in chemical engineering from Northwestern University and his doctorate in polymer science and engineering from the University of Massachusetts at Amherst. Among the many honors Tirrell has received, he was awarded the Polymer Physics Prize of the American Physical Society and was elected into the National Academy of Engineering and the American Academy of Arts and Sciences.

Chairman WEBER. Thank you, Dr. Tirrell.
Dr. Locascio, you're recognized for five minutes.

**TESTIMONY OF DR. LAURIE LOCASCIO,
ACTING ASSOCIATE DIRECTOR
FOR LABORATORY PROGRAMS AND DIRECTOR,
MATERIAL MEASUREMENT LABORATORY,
NATIONAL INSTITUTE OF STANDARDS AND TECHNOLOGY**

Dr. LOCASCIO. Thank you. Chairman Weber, Chairwoman Comstock, Ranking Members Lipinski and Veasey, and Members of the Committees, thank you for the opportunity to discuss NIST's role in enabling advances in materials that strengthen U.S. innovation and industrial competitiveness.

NIST has helped entire industries overcome intractable challenges by measuring materials with ever-increasing precision and characterizing new materials for the very first time. We help American manufacturers be more competitive by enabling development and testing of materials that perform far better than previous generations.

Great leaps in our quality of life are linked to great links in the performance of materials. For example, prosthetics and medical implants, once limited to ceramic and steel and harvested bone, are now made from titanium and polymers and composites. They are stronger, lighter and more functional, helping more people return to work and live active lives.

NIST has been an essential partner to industry in supporting the traditional approach to materials discovery. For example, we have helped the U.S. semiconductor industry, which generates $166 billion in global sales, overcome measurement and material limits to making the smaller, faster chips that the market demands. But traditional materials discovery requires costly trial-and-error cycles. In a new paradigm, NIST supports the use of data and models to simulate materials and predict their performance before spending the money to make them. This approach is called materials by design. GE used materials by design to make new alloys for jet engines in nine years instead of the typical 15 to 20, and the metal in Apple watches was developed and deployed to market in just two years using this approach.

Materials by design is such a game changer that it became a national priority in 2011 with the Materials Genome Initiative. The MGI, as it is known, benefits nearly all economic sectors from the chemical industry to electronics, communications, and biotechnology. The MGI is a partnership among 18 federal agencies, including some in the Department of Energy and Defense, along with NASA and NIST.

NIST supports the MGI with new modeling and experimental capabilities, along with materials data. For example, the Materials Resource Registry is like an online Yellow Pages for materials by design, enabling in-depth, worldwide searches of data collections, computational services, and modeling software. In this registry, we collect and harvest public data from materials science programs in universities, industries, and government to create a valuable national resource, and with access to all this shared data, researchers can more quickly design unique materials for the next great American breakthrough.

To help create an ecosystem for MGI, NIST founded the Center for Hierarchal Materials Design, or ChiMaD, a consortium led by Northwestern University, the University of Chicago, and Argonne National Lab. ChiMaD and NIST together are building tools to support the MGI nationally while advancing technologies that the institute cares about, like 2D electronics and more efficient jet engines. Thanks to the support of Congress, materials by design is gaining ground across the entire U.S. materials science enterprise.

Why is an agency like NIST doing this work? We see ourselves as industry's national lab, a well-respected, trusted, non-regulatory scientific agency that forms strong partnerships with industry to tackle critical national needs. Other countries are investing in their own MGI-like initiatives. The U.S. faces ever-increasing competition in this space. We are still the ones to beat, but we need continued coordination and support among all the players across many sectors to retain this lead.

We greatly appreciate the Members of these Committees and others in Congress for the support of federal acceleration of the innovations in materials science that keep our nation globally competitive and secure and contribute to our quality of life.

I will be pleased to answer any questions you may have. Thank you.

[The prepared statement of Dr. Locascio follows:]

Testimony of

Laurie E. Locascio, Ph.D.
Director
Material Measurement Laboratory
Acting Associate Director of Laboratory Programs
National Institute of Standards and Technology
United States Department of Commerce

Before the

Committee on Science, Space, and Technology
Subcommittee on Energy
Subcommittee on Research and Technology

United States House of Representatives

"Material Science: Building the Future"

June 28, 2017

Introduction

Chairwoman Comstock, Chairman Weber, Vice Chairmen Abraham and Knight, Ranking Members Lipinski and Veasy, and Members of the Committees, I am Dr. Laurie Locascio, Director of the Material Measurement Laboratory (MML) and Acting Director of Laboratory Programs at the Department of Commerce's National Institute of Standards and Technology (NIST). Thank you for the opportunity to appear before you today to discuss NIST's role in and programs focused on enabling fundamental advances in materials that strengthen U.S. innovation and industrial competitiveness. The NIST laboratory programs work at the frontiers of measurement science to ensure that the U.S. system of measurements is firmly grounded in sound scientific and technical principles. Today, the NIST laboratories address increasingly complex measurement challenges, ranging from the very small (nanoscale devices) to the very large (vehicles and buildings), and from the physical (renewable energy sources) to the virtual (cybersecurity and cloud computing). As new technologies are developed and evolve, NIST's measurement research, standards, and services remain central to innovation, productivity, trade, and public safety.

NIST and Materials Science

Many examples of NIST's work I just mentioned are manufactured from materials like steel, cement, plastics, carbon, and silicon. NIST has a role in ensuring accurate measurements of these materials and many more. Why do measurements matter? Measurements give us a common language for the performance of materials and help us have confidence in them, which is particularly important for buildings, bridges, jet engines, and medical devices, and for acceptance of new technologies like wearable electronics and tissue engineering. Since it was founded as the National Bureau of Standards in 1901, NIST has been finding new ways of measuring materials with ever increasing precision and accuracy, and developing ways of characterizing novel materials for the first time. We have world-leading measurement capabilities, in both expertise and equipment, that no individual company or even industry group could amass, and the authorities granted NIST by Congress enable us to respond to the needs of American companies across the manufacturing landscape, working on intractable problems to the benefit of entire sectors. We collaborate closely with other government agencies, including the Department of Energy, and I will highlight some of that work today.

Let me share some examples of NIST responding to the needs of entire sectors. The U.S. semiconductor industry generated global sales of $166 billion in 2015 and held 50 percent of the worldwide market share,[1] but faces increasing competition from overseas. The industry often asks NIST to help them overcome measurement and material limits to making the advanced chips found in the electronic devices used by consumers, scientists and the military. As the semiconductor components in chips become smaller, new materials and processes are required to fabricate these tiny structures, from the substances used to etch the circuits to methods used to automate the assembly of nanoscale components. The semiconductor industry needed new ways to measure novel nanometer-sized structures with sub-nanometer resolution. In response, NIST scientists developed a new, nondestructive way to measure the shapes of semiconductor patterns in three dimensions with X-rays. This method is now used to "see" how polymers can actually assemble themselves into the nanoscale structures needed for semiconductor manufacturing. This method requires minimal retooling, working with the

---

[1] https://www.selectusa.gov/semiconductors-industry-united-states

equipment that chip manufacturers already have in place. To accelerate the transfer of this technology, NIST holds courses to train semiconductor industry chip makers and equipment manufacturers on how to use the method.

Another bedrock of the U.S. economy, the auto industry, has asked for our help in adopting new aluminum alloys, high-strength steels, polymer composites, carbon fibers, and other materials while avoiding traditional—and costly—trial-and-error build cycles. At the NIST Center for Automotive Lightweighting, we measure the micro-scale changes in metals that result from manufacturing processes to illuminate the ideal methods for shaping lightweight alloys while maintaining their resistance to impacts. The NIST Center for Automotive Lightweighting has more than 20 industry partners, including Chrysler, Ford, Novelis, GM, Auto/Steel Partnership, and Alcoa. One major manufacturer used NIST's data to save nearly four months of development time in making a new vehicle that is 700 lbs. lighter and 50 percent more efficient. The sector's scientists have speculated that reducing the number of trail-and-error cycles by half could have significant savings for U.S.-based automakers, [2] and NIST is leading that effort through development of precise measurements. This NIST center collaborates closely with Lightweight Innovations for Tomorrow (LIFT), a Department of Defense-funded institute in the Manufacturing USA network and a recipient of a grant from the Department of Energy.

Nascent industries also come to NIST for help. Additive manufacturing, also called 3D printing, is relatively new, and there is little knowledge and data to ensure that reliable parts can be made on a large scale. An economic analysis estimates that we can lower the costs of additively manufactured parts by 18 percent and save the sector more than $4 billion each year by meeting key technical needs.[3] NIST has led and participated in several industry-driven roadmaps to help identify those key technical needs, and is applying measurement science to every step of the additive manufacturing process so that manufacturers can choose the right starting materials, equipment, and printing and finishing processes to reliably and reproducibly get the properties they want. In addition, we recently launched the NIST Additive Manufacturing Benchmark Test Series (AM-Bench), with help from the Department of Energy and other Federal agencies. AM-Bench engages multiple companies to build the same part using the same starting materials, while multiple computer modelers try to predict the finished parts' shape and physical properties, which NIST material science experts will measure. The results will be available to any company to compare how their own modelling efforts perform against NIST's data, providing them with a way to validate their additive manufacturing simulations and processes. More than forty companies, national laboratories, and universities participate in AM-Bench. New builds, predictions, and tests will recur every two years. The robust simulation tools that result from AM-Bench will replace much of the current trial-and-error approach to making parts that perform as needed.

NIST has a long history of using test series and challenges to bring a community together to solve ambitious problems in support of the NIST mission "to promote U.S. innovation and industrial competitiveness by advancing measurement science, standards, and technology in ways that enhance economic security and improve our quality of life." In 2015 NIST announced its first prize

---

[2] Personal communication with Timothy Foecke, Director, NIST Center for Automotive Lightweighting, at automotive industry project meeting, 2013.

[3] NIST Economic Analysis Brief, "The Economic Impact of Technology Infrastructure for Advanced Manufacturing: An Overview," Gary Anderson, October 2016. http://nvlpubs.nist.gov/nistpubs/eab/NIST.EAB.1.pdf

competition[4]— the Head Health Challenge III in partnership with the NFL, GE, and Under Armour, in which our test results helped to determine a winning material that will garner a $500,000 grand prize. The goal of the challenge was to spur the creation of innovative impact absorbing materials that will result in increased protection for rec-league and professional athletes, warfighters, and first responders. We set up a national prize competition to attract attention from a hugely diverse set of communities, from aerospace to automotive to sports medicine, inviting them to submit their materials to NIST for extensive testing in a broad range of conditions, including temperature extremes from freezing to a hot summer day, and a year's worth of repeated impacts. Participants in two rounds of competition received NIST's test data to further their products; expert judges chose a winner—to be announced in a few weeks—from five finalists. Among the final entries, we saw dramatic technical advances in how well a material can absorb impact. Some of the materials we tested reduced the force of an impact by up to 80 percent, compared to conventional materials. These types of innovative materials can catalyze the development of protective equipment that will mitigate impact and injury in high school athletes or soldiers. In addition to the prospect of a monetary prize, finalists reported that they are using NIST's data to adjust their materials to expand into new product lines or markets, and attract commercial partners for further development and testing.

NIST also provides industry, academia, and other government agencies with access to unique, world-class user facilities that support innovation in materials development. The NIST Center for Neutron Research (NCNR) provides neutron measurement capabilities to the U.S. research community, and the NIST Center for Nanoscale Science and Technology supports the U.S. nanotechnology enterprise from discovery to production by providing access to measurement and fabrication methods and technology. A prime example of the impact that these facilities can have on materials innovation is the work taking place at the NCNR as part of the nSoft Consortium. The nSoft Consortium brings together manufacturers of soft materials (like polymers and biomaterials) and provides a new model to increase their access to the unique instrumentation available at the NCNR. With training and support from NIST technical staff, scientists from companies like Amgen, Dow, and Procter and Gamble use the NCNR facility to measure and characterize new materials in development. This collaboration provides a direct benefit to the industry partners, and helps NIST learn about critical problems in multiple industry sectors so we can develop new capabilities in response. For example, nSoft and its partners developed new sample environments to simulate the extreme pressures, temperatures, and flow rates needed by the pharmaceutical and shale gas industries to measure the structure of fluids in very small geometries. To date, nSoft has helped companies develop therapeutics with a longer shelf-life and higher strength materials, and provided key insight into how plants can stay hydrated in arid and harsh environments.

NIST also partners with a Department of Energy user facility, the National Synchrotron Light Source II at Brookhaven National Laboratory, which produces X-rays that are 10,000 times brighter than the facility it replaces, allowing researchers to see individual atoms in materials. The National Synchrotron Light Source II contributes to the development of new semiconductors for computers and other applications, batteries and solar and fuel cells, superconducting materials, catalysts for chemical production, and materials that can assemble into complex structures by themselves, mimicking how cells, bones, and tissues grow.

---

[4] NIST has a long history of using challenges to engage the public in its many research efforts. Head Health Challenge III was NIST's first cash prize competition announced under the authority of 15 USC § 3719.

These examples show how NIST helps American industries gain competitive advantages by enabling them to more rapidly develop and use new materials with amazing properties: These materials are lighter, stronger, have more function, and require less energy to make or generate less waste in processing. Consumers benefit, in turn, from access to improved products with more features. In fact, historical leaps in our quality of life are linked to revolutionary leaps that improved the performance of materials. Advancing from iron to steel transformed transportation and our built environment. Medical implants—once limited to ceramic, steel, or harvested bone—are now often made from titanium alloys or polymers. They are stronger, lighter, less likely to cause immune reactions, and can be custom made for an exact fit. Your 6-ounce cellphone contains more processing power than a desk-sized 1980's era supercomputer, thanks to innovations in the materials in its integrated circuits, and its usability is greatly enhanced by developments in materials to improve the screen, and the metal case. Many of these advances could not even have been possible without NIST innovations in materials measurements.

As exciting as these advances have been, however, the slow pace at which new and useful materials emerge by trial and error has not changed much since the 19th century. That 19th century approach is inadequate for the 21st century substances that we desperately need to make better prosthetics for injured soldiers; water sensors that detect biological threats and filters that eliminate them; and more efficient solar cells. Typically, it can take 20 years or longer and tens of millions of dollars to find, fine-tune and deploy new a material through a series of trial-and-error experiments. Our Nation's problems are too big to wait that long, and we risk losing America's manufacturing prowess to competition from abroad.

A Materials Innovation Infrastructure

Fortunately, there's a new paradigm for advancing the materials sciences. NIST is a leader in establishing a nation-wide infrastructure so that members of industry, academia, and other government agencies can develop materials *by starting with the properties they need from those materials*, rather than seek to discover materials that might or might not work. This new approach is known as "materials by design." Researchers develop new materials more quickly by working smarter and faster: They use data on the known properties of materials, along with computer modelling, to make informed choices about how to combine or process substances to get the performance they need. GE used this method to make new alloys for jet engines in nine years instead of the usual 15 years. The metals used in Apple watches were developed and deployed to market in just about two years thanks to this approach. Making materials by design is such a game-changer that adoption of the method became a national priority with the Materials Genome Initiative, or MGI, launched in 2011. The MGI launched a partnership among 18 Federal agencies, including the Department of Energy, the Department of Defense, and NIST, that is still very much active today.

Over the years, NIST has conducted millions of measurements to determine the properties of materials, so you can imagine that we have become experts at handling the large amounts of data those measurements generate. That is why NIST supports the MGI with an infrastructure for materials property data. There's no lack of data on metals, polymers, and ceramics: Many research and design programs in the military and government agencies, and in universities and industry, are generating and storing data as well. Making that wealth of knowledge widely accessible as a national resource, however, requires new protocols to ensure that data can be found and is in a recognizable form, and that there are methods for assessing whether the data is of sufficient quality to be useful.

An effort like the MGI requires a variety of industries and industry players, normally fierce competitors, to come together and collaborate effectively—a tall order. NIST has a track record of success in bringing together industry, academia, and government to achieve foundational technological advances, on the strength of its credibility as a trusted, objective third party, concerned with accelerating innovation across entire industry sectors. By developing tools to make better use of the country's wealth of data, NIST combines its own world-class technical expertise with its role as a convener to accelerate innovation across an even larger swath: *all* of the enterprises based in *all* of the materials sciences. We are fostering the availability of information across organizations, locations, and disciplines.

The Materials Development Ecosystem

NIST's position as a well-respected, non-regulatory scientific agency known for objectivity gives us the capability to work with industry and academia to build an infrastructure that enables data sharing. Establishing this data sharing infrastructure amongst competitors requires significant engagement in partnerships that clearly demonstrate how these new approaches will help accelerate innovation.

To help establish this infrastructure and provide a powerful proof of concept, NIST established the Center for Hierarchical Materials Design, a consortium led by Northwestern University, the University of Chicago, and Argonne National Laboratory. ChiMaD, as it is known, is developing the next generation of computational tools, databases, and experimental techniques—an entire ecosystem—to enable the accelerated design of novel materials and their integration to industry. CHiMaD works in emerging areas such as two dimensional electronic materials for advanced computing, additive manufacturing, super-alloys for aerospace engines, and flexible, organic solar cells—programs informed by the many industry members who serve as partners, collaborators, and advisors. ChiMaD also educates undergraduate and graduate students in the materials-by-design approach. Partnerships with Fayetteville State University, a minority-serving institution, and ASM International, a professional society, help share the program with even more communities, preparing a new generation of materials scientists and engineers to meet the workforce needs required by wider adoption of the materials-by-design approach.

ChiMaD hosts the NIST-funded Materials Data Facility, a cloud-based repository where materials scientists can publish, preserve, share, and collaboratively analyze research data. Additional NIST resources in the materials-by-design ecosystem include:

- The NIST Materials Data Repository, a public-access databank hosting about 133 gigabytes of data from 123 groups
- The Materials Resource Registry, which is like a yellow-pages for the materials-by-design approach that enables in-depth, world-wide searches of available resources such as organizations, data collections, data and computational services, and analysis and modeling software
- The High-Throughput Experimental Materials Science Virtual Laboratory, a collaboration with the National Renewable Energy Laboratory to develop a network of computational tools and data resources for comprehensive materials discovery efforts
- The Materials Data Curation System, which helps materials researchers annotate, organize, save, later retrieve, analyze, and share data

Conclusion

NIST has a long history of addressing industry needs with measurement science. Established and emerging industries alike seek our measurements and methods to help them innovate and remain competitive globally, contributing to our economy. While traditional materials science research has been trial and error, the demand for accelerated innovation requires a new approach. NIST is establishing the necessary data and computational infrastructure that allows designers to intentionally develop new materials for the properties their new products require. In these efforts, NIST is proud to be regarded, as it has for more than a century, as a trusted, neutral third party that facilitates collaboration among industry, academia, and government agencies to meet critical national needs.

We greatly appreciate the efforts of the members of these committees and other members of Congress to support Federal agency acceleration of the innovations in the materials sciences that keep the Nation globally competitive and secure, and that contribute to our quality of life.

I will be pleased to answer any questions you may have.

Laurie E. Locascio

Dr. Laurie E. Locascio is the Acting Associate Director for Laboratory Programs (ADLP) at the National Institute of Standards and Technology (NIST). As Acting ADLP, she provides direction and operational guidance for NIST's scientific and technical mission-focused laboratory programs and serves as principal deputy to the Under Secretary of Commerce for Standards and Technology and NIST director, among other duties.

Dr. Locascio's current permanent position is director of the Material Measurement Laboratory (MML) at NIST. MML, one of seven research laboratories within NIST, has an annual budget of $175 million and nearly 1,000 federal employees and guest researchers from industry, universities, and foreign laboratories. MML provides a measurement science and standards infrastructure for the Nation's industries based in the biological, chemical and materials sciences, promoting U.S. innovation and industrial competitiveness in ways that enhance economic security and improve our quality of life. MML is a source of unbiased measurement standards, data, and cutting-edge methods and technologies that promote innovation, market readiness, and quality control in vital economic sectors.

MML develops measurement standards in the form of documented measurement methods and instrument calibrations, and coordinates the NIST-wide Standard Reference Material® and Standard Reference Data programs. MML provides more than 1,200 Standard Reference Materials that ensure the accuracy of millions of measurements vital for efficient manufacturing, acceptance of American-made goods in international markets, regulatory approval of new technologies and medical treatments, and consumer confidence.

Dr. Locascio previously served as chief of the Biochemical Sciences Division in the Material Measurement Laboratory. She has published more than 100 scientific papers and holds 11 patents in the fields of microfluidics, biosensors and sensor/flow systems. She is a Fellow of the American Chemical Society (ACS) and the American Institute for Medical and Biological Engineering (AIMBE). She received her doctorate in toxicology from the University of Maryland at Baltimore.

Education

Ph.D. in toxicology from the University of Maryland at Baltimore

M.Sc. in bioengineering from the University of Utah

B.Sc. in chemistry from James Madison University

Chairman WEBER. Thank you, Doctor.
Dr. Schwartz, you're recognized for five minutes.

TESTIMONY OF DR. ADAM SCHWARTZ, DIRECTOR, AMES LABORATORY

Dr. SCHWARTZ. Good morning. Chairman Weber, Chairwoman Comstock, Ranking Members Veasey and Lipinski, and Members of the Committee, thank you for the opportunity to testify at this hearing, and thank you for your continued strong support of materials research.

The United States is the world leader in materials science, condensed-matter physics, and chemistry research. Federally funded research has created an innovation system unmatched anywhere including the private sector. Our leadership is due in large part to governmental science funding across the continuum from grand challenge and use-inspired basic research to applied research and technology deployment. As a country, we've reaped tremendous benefits in the economics, energy security, national security, and our quality of living. The United States leads in discovering and applying materials with novel properties.

New materials discoveries enabled by basic research at our national laboratories and universities have significant economic and societal impacts on our everyday lives. Consider your smart phone, tablet or almost any other consumer electronic device. Ames National Laboratory and Sandi National Laboratories collaborated to create a lead-free, environmentally friendly replacement for lead-based solder. This advanced alloy was ultimately licensed to over 65 companies in 23 countries with an economic benefit to the private sector estimated at $610 million per year.

New and experimental—new experimental and computational capabilities developed from sustained federal investment in a talented and dedicated scientific workforce have accelerated the pace of discovery of novel materials. We can now design and create materials tailored for some specific purposes and soon will be able to do so much more broadly if appropriate research continues.

Great opportunities abound for new materials to impact our world. LED lighting transformed a century-old light bulb industry that hadn't advanced since Edison. Research to replace the current 100-year-old compressed-vapor refrigeration with solid-state magnetic technologies enabled by new materials could potentially reduce our energy consumption by one-quarter and have transformative impacts.

An amazing opportunity also exists in information technology. For decades in the computer industry, the density, speed and computational power of integrated circuits have increased exponentially over time as predicted by Moore's Law but we're fasting approaching the theoretical limits of processor materials. To go beyond Moore Computing, research is needed to create new quantum materials that use less energy and provide computing power beyond today's approaches with conventional silicon chips.

Tremendous opportunities exist in additive manufacturing, or 3D printing of metals to fabricate parts for the military, biomedical, and aerospace industries. Currently, progress is constrained by a lack of fundamental understanding and control of kinetic processes as well as a lack of suitable metal powders. Collaborations between Ames and other laboratories are pooling their expertise to meet these needs, establishing U.S. leadership in a fast-growing industry.

The biggest challenge facing U.S. materials research right now is maintaining our global competitive edge. The rest of the world is catching up. Countries like China, South Korea and India are investing increasing percentages of their GDP in materials research and our global competitive advantage in this key enabling science is under threat. Will the United States be the first to invent the next catalyst and in a $30 billion petrochemical industry, discover the material that will replace traditional semiconductors in the $350 billion electronics industry, or provide options for the next critical material on which our military systems depend? The private sector cannot do this by itself.

Federally funded research enables world-changing materials advances like the ability to address critical material shortages through the basic research provided by the Critical Materials Institute and the ability to design and create new materials to revolutionize the electronics, lighting, refrigeration and air conditioning industries, among many other manufacturing sectors. The key to future success is sustained research on fundamental principles and the resulting discovery of advanced materials.

Ames Laboratory, like other national laboratories and research universities, is on the cusp of great materials discoveries that will further the nation's economic, energy and national security interests but we need your continued support and resources to meet our mission.

Thank you for the opportunity to testify today, and again, thank you for your consistent support of materials research. This Committee's leadership has paved the way for remarkable innovations. I'd be happy to address any questions or provide additional information.

[The prepared statement of Mr. Schwartz follows:]

Testimony of
Adam J. Schwartz
Laboratory Director, U.S. Department of Energy Ames Laboratory
Professor of Materials Science and Engineering, Iowa State University
Ames, IA 50011
Before the
Joint Subcommittee on Energy and Subcommittee on Research and Technology
of the
Committee on Science, Space, and Technology
U.S. House of Representatives
Materials Science: Building the Future
June 28, 2017

Chairman Weber, Chairwoman Comstock, Ranking Members Veasey and Lipinski, and Members of the subcommittees, thank you for the opportunity to testify at this hearing and thank you for your continued strong support of materials research. I am pleased to have the opportunity to discuss building the future of materials science. I am Adam Schwartz, director of Ames Laboratory, a Department of Energy national laboratory located on the campus of Iowa State University, in Ames, Iowa. Ames Laboratory is a single program Office of Science laboratory with the mission to create materials, inspire minds to solve problems, and address global challenges. I am also a professor of Materials Science and Engineering in the College of Engineering at Iowa State University. Before arriving at Ames Laboratory 3 years ago, I was division director for the Condensed Matter and Materials Division, acting program leader for the Dynamic Properties of Materials and the Physics and Engineering Models Programs, and led the plutonium aging program during my nearly 23 years at Lawrence Livermore National Laboratory, a National Nuclear Security Administration national laboratory in Livermore, CA.

The United States is the world leader in materials science, condensed-matter physics, and chemistry research. Federal investment in academia, national laboratories, and scientific user facilities—core materials science research supported by both the Department of Energy and the National Science Foundation—has created an innovation system that is unmatched anywhere in the world, including the private sector. Our leadership is due in large part to materials science funding across the continuum, from grand challenge and use-inspired basic research to applied research and technology deployment. As a result, the U.S. leads in discovering and applying new materials with novel properties. As a country, we've reaped tremendous benefits in economics, energy security, national security, and our quality of living.

The Department of Energy national laboratory system created after the end of the Manhattan Project successfully built upon and greatly expanded materials research and development expertise that was critical to the Project's success. Today, this unparalleled expertise remains unmatched and our country's national laboratories constitute their own unique facilities and research strengths, and working together to contribute significant technological advances in the national interest. Fundamental and highly specialized expertise in the synthesis of materials; theory, modeling and simulation; as well as materials characterization and development across the national laboratory network are collaboratively brought to bear on solutions to complex scientific challenges. Ames Laboratory and other national laboratories breed materials innovations that over and over again lead to revolutionary technologies.

Materials Discoveries Have Impact: Lead-free Solder

At the close of the 20th century, the production of electronics boomed as personal electronic devices such as computers and mobile phones became staple household items. As the production of electronics expanded, so did concerns about electronic waste and its potentially harmful environmental effects, particularly with the use of leaded solders required to fuse electronic components together. In the mid-1990s, the electronics industry set a goal of halting the use of lead-based solders due to their harmful effects on people and the environment. Research to find a lead-free solder increased in response to this goal.

Lead-free solder, developed at the Ames Laboratory in collaboration with Sandia National Laboratory, is a tin, silver, and copper alloy that is low melting and applies easily, at a reasonable cost. This revolutionary solder alloy replaces many uses of the traditional tin-lead low melting solder, reducing further the number of lead toxicity hazards in our everyday environment. This advanced alloy was ultimately licensed to over 65 companies in 23 countries with an economic benefit to the private sector estimated at $610M per year and is a key component in manufacturing mobile phones, tablets, and almost every consumer electronic device. It has generated approximately $60 million in royalty income, making it the top all-time royalty generator for Ames Laboratory and Iowa State University and one of the top for the U.S. Department of Energy. It is the only invention in the national laboratory complex that has returned revenue to the U.S. Treasury, approximately $6M.

Materials Discoveries Have Impact: Spinoff companies
Basic research discoveries at Iowa State University and Ames Laboratory have resulted in spinoff companies that demonstrate the power of federally funded research across the continuum of science. Research at Iowa State University and Ames Laboratory found that co-locating catalysts and their supports in confined spaces greatly increases reaction rates. That discovery led to a new series of bi-functional mixed metal oxide materials for a new type of catalysis - cooperative catalysis. Further research identified these new materials as ideal for biodiesel production. Starting with funding from DOE Basic Energy Sciences, then from Energy Efficiency and Renewable Energy, this new material attracted investor capital and Catilin, Inc. was born. A number of years later, a U.S. multi-billion dollar company bought Catilin and brought the catalyst to market.

For nearly 30 years, Ames Laboratory has also been a pioneer in the development of metallic powders, first for use in powder metallurgy, and now for use in 3D printing. Advances in the fundamental understanding of powder processing led to numerous patents and ultimately the spinoff of Iowa Powder Atomization Technologies, IPAT, with the goal to create high purity titanium metal powders that can be formed into industrial parts for military, biomedical, and aerospace applications. Using gas atomization nozzles and pour tubes developed at Ames Laboratory, the titanium powder making process is more efficient than anything known in the past and, thus, lowers the cost of the powder. IPAT was acquired three years ago by a different U.S. multi-billion dollar company and now makes commercially available titanium powders.

Great Opportunities Abound in Materials Research

Just like lead-free solder, cooperative catalysts, and metallic powders, great future opportunities exist to impact our world. New experimental and computational capabilities, developed from sustained federal investment in a talented and dedicated scientific workforce, have accelerated the pace of discovery of novel materials and changed the lives of every American for the better. We can now design and create materials tailored for some specific purposes and soon will be able to do so much more broadly with appropriate research support. The time is right to grow this capability because the nation's economic, energy, and security future relies on the discovery of advanced materials for efficient energy conversion, generation, and transmission.

Refrigeration and air conditioning

Modern civilization is highly dependent on reliable refrigeration and air conditioning and yet, we still use a vapor-compression technology that has remained essentially unchanged for over 100 years to provide vital societal needs ranging from food supply to medical treatment. Just like fluorescent and LED lighting transformed a century old industry that hadn't advanced since Edison's incandescent bulb, research of new materials to replace current compressed-vapor refrigeration technology could potentially reduce our energy consumption by one quarter.

Over the past century, all parts of a conventional refrigerator were refined due to concerted research and development efforts. Future improvements, however, may only be incremental since vapor-compression refrigeration is already near its fundamental limit of energy efficiency. Yet residential and commercial cooling still consumes one out of every five (or more) kilowatt-hours of electricity generated in the U.S.

New technologies with a potential to save a substantial fraction of the estimated 20 to 25% of the generated electricity used for cooling, refrigeration, and air conditioning in the U.S. will make a tremendous impact on our nation's energy future and ensure lasting business competitiveness and leadership in the industry, which is projected to double over the next 5-7 years to over $120B annually. Caloric cooling is a revolutionary, early-stage, solid-refrigerant technology making use of quantum effects that approach ultimate efficiency. Numerous system-level studies have predicted as much as 20 to 30% higher electrical energy efficiency for caloric-based refrigerators compared to the conventional vapor-compression systems. Solid-state caloric cooling today is analogous to where conventional refrigeration was in the 1920s, at the edge of a major breakthrough relying on the discovery and development of advanced materials.

Caloricool™ was established in 2016 as a DOE Energy Materials Network Consortium led by Ames Laboratory, as a collaboration of national laboratories, universities, and industry, to discover critically needed high-performance solid refrigerant materials that are earth-abundant, safe, manufacturable, and affordable. The consortium seeks novel materials that will meet or exceed the performance of conventional coolants and can be accepted by U.S. manufacturers for deployment within five to ten years.

Information technology

Amazing opportunities also exist in information technology. For decades, the computer industry has operated under Moore's Law, which predicts the density, speed, and power of integrated circuits increases exponentially over time. But the state-of-the-art in computing power is rapidly approaching theoretical limits of the materials that make up the heart of our computers and processors.

While increases in computational power have followed Moore's Law, the limitations of materials are causing an increase in energy costs for modern scientific computers that eventually will make supercomputers impractical because of enormous energy consumption and the need to dissipate massive amounts of heat. The density of transistors has approached atomic spacing, speed has plateaued, and the energy required to run systems is rising sharply. Industry can't indefinitely continue to cram more transistors onto chips.

To go beyond Moore Computing, research is needed to create new quantum materials and chips that use much less electricity, or even energy other than electricity, for example magnetic field or light, to control transistors and provide computing power and support communications far beyond today's approaches with conventional silicon chips. Targeted and well-focused fundamental research in materials and materials structures underpins the development of these new frontiers.

The goal is to discover and develop materials that enable very low power consumption per instruction. In order to realize this goal, new materials and materials structures controlled with low fields are required. Independent of the particular materials structure or structures, there are fundamental problems that cut across. First, how do we move electrons at nanometer length scales and make them respond at picosecond (one trillionth of a second) timescales in inhomogeneous nanoscale structures? Next, how do we ensure thermal stability? Third is how do we develop materials that will enable low energy transistors? Finally, how do we synthesize and integrate new materials into manufacturable architectures?

No suitable materials solutions exist today, but there are various potential materials systems that form a basis for both near and long-range roadmaps, and opportunities for expanding materials science understanding and consequent development efforts. Materials that are more developed for information technology applications and, therefore, could yield near term partial solutions include superconducting materials, organic materials, magnetic semiconductors, and materials for spintronics logic. Materials in their more nascent stage of understanding and development, but that could ultimately provide superior performance, include those that can be switched at will between conducting and insulating states and topological matter.

Critical Materials

Throughout the history of humanity, every civilization, including ours, revolved around materials that were discovered and put to use at the time. If any one of the materials required to make a product in high demand is suddenly no longer available, or in other words becomes critical, life undergoes changes that in the past were often catastrophic, unless a society foresees and addresses the shortage ahead of time.

The Critical Materials Institute, a DOE Energy Innovation Hub established in 2013, is conducting early stage research to accelerate the search for and establish alternatives to rare earth materials for a large range of advanced technologies, both reducing environmental impact and the nation's dependence on sensitive foreign sources. The consortium's team is comprised of national laboratories, universities, and industrial partners.

Rare earth elements are the most prominent of the critical materials today. CMI aims to develop economically viable processing techniques for improved availability of critical materials for clean-energy technologies, develop new techniques to recover them from waste and scrap, and find acceptable alternatives for use in devices such as generators, motors, lighting, and magnets.

Based on early stage foundational research, CMI has filed 33 U.S. patent applications and 62 records of invention that address rare earth separations for primary metal refining or recycling processes and new magnet compositions and processes. For example, CMI has discovered suitable green and red lamp phosphor substitutes, reducing the use of the rare earth terbium by 90% and eliminating the use of the rare earth lanthanum for green phosphors, and eliminating the use of rare earths europium and ytterbium in red phosphors. Industry is currently assessing the feasibility for commercial lighting via full manufacturing trials.

Additive Manufacturing

There is also tremendous opportunity for U.S. leadership through advances in additive manufacturing or 3D printing of metals, which builds components from the bottom-up by selectively depositing melted metals layer-by-layer. Intense industry interest is focused on additive manufacturing of metal alloys based on a host of potential benefits and applications, including design development, prototyping, customization, and reduction of production waste and costs.

Because of these benefits over traditional manufacturing methods, additive manufacturing processes promise to initiate a renaissance in American manufacturing. Based upon these potential widespread impacts, some expect the economic output of the additive manufacturing industries to be at least $3.1B in 2016 and $5.2B in 2020.

However, a growing consensus within the additive manufacturing community has identified challenges: a lack of fundamental understanding and control of additive manufacturing processes and high quality metal powders necessary to produce the desired properties needed for robust metallic parts. A collaboration between SLAC, Lawrence Livermore National Laboratory, and

Ames Laboratory seeks to provide this important foundational knowledge in manufacturing process control. To overcome the powder-based challenges, Ames Laboratory has established a Powder Synthesis and Development Facility to form a strong partnership with the research community affiliated with the Manufacturing Demonstration Facility at Oak Ridge National Laboratory. This early stage work complements current investments being made by industry, the government, and academic institutions, and the broad return on the investment into this research will be significant.

Summary and Conclusions

The biggest challenge facing U.S. materials research right now is maintaining its global competitive edge. The rest of the world is catching up. Countries like China, South Korea, and India are investing increasing percentages of their GDP in materials research and our competitive advantage in this key enabling science is under threat. Will the U.S. be the first to invent the next catalyst in a $30B petrochemical industry, discover the material that will replace traditional semiconductors in a $350B electronics industry, or provide domestic options for the next critical material on which our military systems depend? The private sector cannot do this by itself. Federally funded research enables world-class materials research, like the ability to address critical material shortages through the basic research provided by the Critical Materials Institute, and the ability to design and create new materials to revolutionize the electronics, lighting, and refrigeration and air conditioning industries, among many other manufacturing sectors. With the current focus on manufacturing and energy to the nation, now is the time to grow and leverage materials research.

Ames Laboratory, like other national laboratories and research universities, is on the cusp of great materials discoveries that will further the nation's economic, energy, and national security interests, but we need your continued support and resources to meet our mission.

Thank you for the opportunity to testify today and again, thank you for your consistent support of materials research. This committee's leadership has paved the way for remarkable innovations. I would be happy to address any questions or provide additional information.

Dr. Adam J. Schwartz is Director of Ames Laboratory, a Department of Energy national laboratory located on the campus of Iowa State University in Ames, Iowa. He is also a professor of Materials Science and Engineering in the College of Engineering at Iowa State University. Ames Laboratory is a single program Office of Science laboratory with the mission to create materials, inspire minds to solve problems, and address global challenges. Prior to joining Ames Laboratory in 2014, he had nearly 23 years of materials science research and management experience at Lawrence Livermore National Laboratory in Livermore, CA that spanned physical metallurgy to condensed matter physics with a particular focus on phase transformations, phase stability, and electronic structure of actinides and lanthanides. He held leadership positions for Plutonium Aging, Dynamic Properties of Materials, and Physics and Engineering Models programs. He led the Lawrence Livermore National Laboratory team during the development of the Critical Materials Energy Innovation Hub then continued on as the Developing Substitutes Focus Area Leader. In this role, he leveraged his expertise in metallurgy, condensed matter physics, and technical management to drive the innovation of substitute materials. Dr. Schwartz has authored over 100 journal articles, monographs, book chapters, technical reports and co-edited two editions of *Electron Backscatter Diffraction in Materials Science* (Kluwer Academic/Plenum Publishers, 2000; Springer, 2009). Dr. Schwartz holds B.S. and M.S. degrees in Metallurgical Engineering and a Ph.D. degree in Materials Science and Engineering from University of Pittsburgh.

Chairman WEBER. Thank you, Dr. Schwartz.
Dr. Higgs, I recognize you for five minutes.

TESTIMONY OF DR. FRED HIGGS, JOHN AND ANN DOERR PROFESSOR OF MECHANICAL ENGINEERING, RICE UNIVERSITY

Dr. HIGGS. Chairman Weber, Chairwoman Comstock, Ranking Members Veasey and Lipinski, and other Committee Members, I appreciate the opportunity to testify before the Subcommittees. As the John and Ann Doerr Professor of Mechanical Engineering and the Faculty Director for the Rice Center for Engineering Leadership at Rice University, I am excited about this opportunity to provide this testimony today.

Today I'm here to discuss the broad economic impact of materials science on the nation and the need for America to invest big in basic science in this area and other fields of engineering, which are catapulted forward by materials advances.

Whenever you see a new flurry of research activity or new radically high-performing technologies, this is almost always related to some type of material advancing or technology deployment that finally figured out how to use a cutting-edge material which was discovered by basic science research no less than a decade ago.

Today I'll discuss new material advancements, science competitions, and industry lab partnerships.

New materials can improve the safety and environmental impact of energy production technologies. In terms of oil and gas drilling, the development of effective, environmentally friendly additives and drilling mud may enable more efficient cooling, lubrication and rock cutting removal from the drill-rock interface. More efficient and environmentally safe extraction processes allow workers to have less exposure to dangerous activities as well. Material advancements can reduce the impact that energy production processes such as coal and natural gas combustion have on our environment.

There are also technological benefits of material advancements in orthopedic medicine. Advanced coating such as nanocrystalline diamond are very robust and compatible with the human body.

There are technological benefits of material advancements in transportation. Tire rolling, resistance and high traction compete to hinder fuel performance. Basic science involving nanomaterials are expected to improve tire performance and are expected to save maybe $35 million barrels of oil annually.

There are technological benefits of material advancements in manufacturing, particularly additive manufacturing, which most here may know as 3D printing, as you heard my predecessor say. More advanced innovations such as composite materials and greater materials remain underdeveloped. 3D printers are also super slow and cannot speed up until fundamental material science questions are answered.

I would like to address another point: crowdsourced-based science prize competitions. One of the new successful strategies for inspiring open innovation and accomplishing idea mining is science prize competitions. While these can be exciting, as my team has competed in them, the potential loss of university IP can in some cases be in danger when the fine print of such competitions re by entering this competition, we can use your ideas without permission whether you win or lose. Normally those are industry-based competitions. The Committee should employ careful oversight of the non-defense agencies' ability to initiate competitions that university researchers perceive as exploitive.

In terms of the merits of university-lab partnerships, government labs serve many noble purposes for our nation from an academic viewpoint. First, they provide our government with research capacity and the personnel and equipment infrastructure to tackle the nation's most pressing problems. Second, they provide a rich research ecosystem of researchers who care about the science of discovery divorced from the pressures of generating quarterly profits, and third, they provide collaborative resources in terms of intellectual capital, equipment and mentorship for young researchers. I work with different agencies and labs such as NASA Glenn and NETL. I can honestly say that just like many of my other colleagues who work with government labs, their support of our research has been pivotal in helping people like me mature from a young professor into a leader in my field. Federal labs have even provided guidance to startup companies such as my own NSF-funded SBIR company, InnovAlgae. DOE labs such as Inrel have advised us of the best path toward technology validation including connecting us to industrial partners that could benefit commercialization efforts.

There are also merits in university-company partnerships. A seasoned researcher at a Fortune 500 company once said to me universities use money to create knowledge but companies use knowledge to generate money, but these days, many companies are desperately looking for Ph.D.'s to hire from universities and yet they spend no money supporting university research. A perfect storm is being set up where companies expect Ph.D.'s to just magically be output without anyone making an investment input. Meanwhile, other countries in Asia and Europe are strategic, creating a Ph.D. investment training and hiring cycle that has catapulting their nations over America, the country I so dearly love. It would be a game changer if companies tax-incentivized to invest seed money into university-based research.

And I leave you with the final recommendation for supporting basic research. If Congress were to inject new funds into NSF to increase the number of graduate fellowships from just a factor of two from 2,000 to 4,000, it would be a big game changer in terms of supporting basic research. Thank you.

[The prepared statement of Dr. Higgs follows:]

The Broad Economic Impact of Material Science Advancements on fields such as Tribology and Particle Technology

WRITTEN SUBMISSION

Testimony before the Subcommittees on Energy and the Subcommittee on Research and Technology,

of the Committee on Science, Space, and Technology,

United States House of Representatives

June 28, 2017

C. Fred Higgs, III, Ph.D.

The Department of Mechanical Engineering, and the Rice Center for Engineering Leadership

Rice University, Houston Texas

I appreciate the opportunity to provide this written testimony to the Committee. I am the John & Ann Doerr Professor of Mechanical Engineering, and the Faculty Director of the Rice Center for Engineering Leadership at Rice University in Houston, Texas.

Material science has had a broad impact on our nation, requiring substantial investments in basic science to catapult forward the promise of engineering and science fields such as mine own, Tribology and Particle Technology.

I have over 100 research publications, almost all of them in the areas of Tribology, the study of materials in sliding contact and the associated friction, lubrication, and wear. Tribology comes from a Greek word meaning "to rub" and this gives you a picture of what Tribologists do. Within the field of Tribology, which is largely a field advanced by material scientists, physicists, and mechanical engineers like me, I study problems where surfaces are rubbing against each other and particles are present. This means that I also have expertise in particle technologies, such as those found in the multi-billion dollar solids processing industries. I have numerous pending and provisional patents and my research has been sponsored by federal agencies and departments such as the National Science Foundation (NSF) and the Department of Energy (DOE), and industry, private foundation, and venture-capital entities. I am passionate about mentoring the next-generation of engineering and technology researchers, having advised 100 undergraduate, 30 Masters, 14 doctoral, and 5 postdoctoral research students over the last 13 years. Some of my research has been translated to industry, in the form of computer modeling approaches, and in high-tech start-up activity, such as InnovAlgae, a university research spin-off company I co-founded, that develops advanced algae-related production technologies for the energy and consumer health industries.

There are three topics relevant to my testimony today:

1. New materials can improve the safety and environmental impact of energy production technologies
2. Material advancements can provide the foundation for new technologies in medicine, transportation, manufacturing and computing.
3. The merits of (i) science prize competitions, (ii) university-federal lab/agency partnerships, and (iii) university-company partnerships, in speeding the development of advanced materials.

A final recommendation for supporting basic science research can be found at the end of this document.

New materials can improve the safety and environmental impact of energy production technologies.

Mining and Drilling: energy resource extraction processes. **Mining** operations for energy resources such as coal, in addition to **drilling** for resources such as natural gas can both benefit from advanced coatings. Not only can lubrication-friendly coatings promote energy efficiency in these sectors which expend a lot of energy, wear-resistant coatings can extend lifetimes of drill bits and mining components, thereby accelerating extraction time. In the drilling industry, 'time is money' to the levels of hundreds of thousands of dollars per day so the economic impact can be positively affected if the drilling times are shortened. The development of effective, environmentally friendly additives in drilling muds, may enable more efficient cooling, lubrication, and rock cuttings removal from the drill bit/rock interface. More efficient and environmentally safe extraction processes allow workers to have less exposure time to dangerous activities and likely ensure there are less disturbances to the environment.

Material advancements can also move forward technology that reduces the impact that fossil fuel energy production processes. Material advancements can reduce the impact that energy production processes such as coal and natural gas combustion have on our environment. For example, chemical looping is a thermal particle combustion process that allows you to efficiently separate the carbon dioxide CO_2 generated from coal combustion into a pure stream that can be easily taken from a power station's flue gas and captured for long-term storage. A key part of this process being as environmentally safe as possible is the development of innovative metal oxide materials[1] to serve as the oxygen carrying solid particles. For example, nickel (Ni) based oxygen carriers are highly effective in chemical looping yet they are expensive and somewhat toxic. Thus, basic science is needed to find viable non-Ni alternatives to move the technology forward. The upside is that a successful chemical looping process could serve as a viable carbon capture and storage technology, helping America to better enjoy the spoils of its vast natural gas finds without it automatically being detrimental to the environment. My point is not to advocate for this technology, but to show that material advancements, which come from basic science, can make powerful technologies with non-favorable environmental impact, more favorable. The NSF is funding projects on chemical looping and such basic science research can end up producing win-win technology and economic scenarios for our country where the environmental impacts of fossil fuels are lessened.

Material advancements can provide the foundation for new technologies in medicine, transportation, manufacturing and computing.

The technological benefits of material advancements in orthopedic medicine. Many Americans have artificial joint implants for their hips or knees. For example, about 250,000 older people age 65 and older are hospitalized for hip fractures each year[2], and almost 95% of these are caused by major falls. By 2030

---

[1] **Source**: *Jing, D., Mattisson, T., Ryden, M., Hallberg, P., Hedayati, A., Van Noyen, J., & Lyngfelt, A. (2013). Innovative oxygen carrier materials for chemical-looping combustion. Energy Procedia, 37, 645-653.*

[2] Centers for Disease Control. National Center for Health Statistics. Health Data Interactive. [Oct. 2015].

alone, the number of hip replacement surgeries in America is expected to explode to 572,000 annually[3]. If you were like my father, who got an artificial knee just a few years ago in his 70s, the surgeon might have told you to wait as long as possible because the implants may only last for 10-15 years. This is because human joints such as knees and hips are synovial joints which means they rub together and have lubricants, similar to mechanical bearings. Unlike the healthy, natural joint, artificial joints are often in partial contact when they are rubbing together. Therefore, these joints wear out and have finite lifetimes so orthopedic surgeons often push their patients to wait as long as possible or until the pain becomes unbearable before getting replacement joints which have finite lifetimes due to the wear and tear.

However, novel materials such as diamond-like carbon (DLC) and nanocrystalline diamond[4] (NCD) may lead to long-lasting artificial joint implants due to their ability to provide the ultimate wear-resistant coating. Advanced coatings such as NCD are not only very smooth when deposited on orthopedic materials, they are hard like diamond, chemically inert (i.e., non-reactive) and compatible with the human body. This type of advance could lead to fewer patients enduring long-term pain (which lessens the need for costly, addictive pain medications) and allows joint replacement surgeries to occur earlier in life – perhaps even before Medicare coverage begins.

The technological benefits of material advancements in transportation. New material technologies will also help advance American transportation. A report[5] commissioned by ARPA-E in 2017 and led by researchers from the University of Pennsylvania in concert with other researchers from government and academia outlined some basic opportunities for materials to advance the transportation sector beyond just finding lighter materials. For example, the tires in automobiles and trucks are both blessings and curses. The tires need to grip the road through robust traction performance, and they should also have long lifetime before they wear out. The rolling resistance from today's tires consume a lot of energy in the form of fuel. In other words, it takes more energy to push your child on a bike with tires than to push them on ice skates. However, ice skates do not brake as well as rubber tires. Thus, the goal is to find better tire material which can both roll and brake easily. Material advances such as nanomaterials, which currently are at Technology Readiness Levels[6] (TRL) as low as 2, could soon make their way from the basic science lab to the applied science lab or commercial company's demonstration floor. Tire rolling resistance and high traction compete to hinder fuel performance, but the problem was alleviated somewhat in the 1990s when silica as a reinforcing nanofiller improved performance. Today, thanks to basic science discovery, we have a much better understanding of nanomaterials. With the appropriate level of funding, it is just a matter of time before material discovery leads to an even better tire technology. And the potential gains are enormous, since these proposed new tires have been estimated to save American up to 0.2 quads per year (i.e., 35 million barrels of oil equivalent per year or 0.2 quadrillion Btu, which equals 0.21×10^{18} J).

---

[3] Stryker Corporation website, 2015.
[4] Source for NCD Bio: Narayan, Roger, ed. *Diamond-based materials for biomedical applications*. Elsevier, 2013.
[5] Source: Carpick, R. W., Jackson, A., Sawyer, W. G., Argibay, N., Lee, P., Pachon, A., & Gresham, R. M. (2016). The tribology opportunities study: can tribology save a quad?. *Tribology & Lubrication Technology, 72*(5), 44.
[6] Source: https://en.wikipedia.org/wiki/Technology_readiness_level

The technological benefits of material advancements in manufacturing. Traditional manufacturing certainly expends a lot of energy and resources. New materials in the form of advanced coatings can lead to economic gains in the manufacturing industry since machining operations often have cutting tools which wear away quickly, expend excessive energy, or thermally damage the part being machined due to improper cooling. However, additive manufacturing (AM) or 3D printing has already been a game-changer, namely in the way companies are designing their future manufacturing strategies to take advantage of the $6B global AM market.

Three-dimensional (3D) printing means there will be a rise of 'mass-customization' instead of the normal paradigm of 'mass manufacturing', where one-size-fits many. AM technologies promise an economic landscape where in many cases, parts can be produced faster, cheaper, yet while expending less energy and wasting less resources. Powder-bed 3D printing consists of the technologies that produce parts or components, slice by slice, by spreading a layer of the desired material in powder form and using different techniques to bind the powder together into a solid 3D final form. While nearly 300,000 consumers have purchases low-cost (sub-$5000) 3D printers, these are not powder-bed 3D printers. The real excitement in terms of next-generation technology is from high-end metal powder-based 3D printers, since many of those can produce direct metal parts. Once you are able to print metal materials, the opportunities for innovating new technologies or re-making old ones are boundless. However, current metal 3D printers print a very limited set of metal materials and usually cannot handle most alloy material systems. While there are numerous impressive-looking metal 3D printers, their ability to print is limited to a small group of metals such as titanium, stainless steel, Inconel (the material from which airplane black boxes are made), aluminum, cobalt chrome, and precious metals such as gold and silver (which is why the 3D printed jewelry industry is exploding).

However, more advanced innovations such as composite materials and graded materials (e.g., a metal sheet that is comprised of two materials which are varied in percentage from one material at one end to the other material at the opposite end). These printers are also super slow and cannot be sped up until fundamental material science questions are answered such as how different powder materials are precisely deformed during sintering, both locally and globally.

The technological benefits of material advancements in computing. Discovered through basic science research over many years, carbon nanotubes (CNTs) are small cylinders whose diameters are about a nanometer in size. They have extraordinary thermal conductivity, mechanical, and electrical properties, and consequently are increasingly being tested as devices in computer chips, such as CNT transistors. A similar carbon related material is graphene, which is 200 times stronger than steel but with high thermal and electrical conductivity. It is also being explored as a new material for computing and many other applications.

Employing novel approaches to speeding the development of advanced materials.

The merit of 'science prize competitions'. I am aware that one of the new successful strategies for inspiring open innovation and accomplishing idea mining is science prize competitions. My laboratory has participated in a few of them, including being minor beneficiaries of one from an Australian company. It should be recognized that sponsors (whether they be a private, third party 'open innovation company' such as Nine Sigma, a large technology company such as Microsoft, or a federal agency such as DARPA) of such competitions are usually the beneficiaries of upside-down cost-benefit ratios, in comparison to what the academic researchers give up.

For example, I participated in a science competition of a very large high-tech company, who offered to give the 'winners' some new hardware technology and $100K for sponsoring their new idea. Upon developing my proposal, I was about to submit it until I read the fine print which said the sponsor company *"has access to, may have or have had possession of, and/or may create or has created materials and ideas which may be similar or identical in format or other respects. I agree that I will not be entitled to any compensation because of the user of any such similar or identical material ..."*

Thus, had I submitted my ideas and not won, the company may still have used it without any expectation of compensation for me or my employer. I hypothesized that they awarded about 5 of 200 submissions, which means they only needed to invest a few hundred thousand dollars for the benefit of an enormous amount of university IP. I question if this is a cunning violation of the Bayh-Dole act which has spawn so much economic success since the 1980s, including some of the most famous university start-up companies (e.g., Google,) around today.

The potential loss of university IP (and the revenues it brings back to the school) are in danger since many academics participate in the current low-odds of winning, basic science research funding environment. I believe the committee should employ careful oversight of non-defense agencies' ability to initiate competitions that university researchers perceive as exploitive. If the sponsor company I encountered had said that they would data mine the submissions for new ideas and contact the proposers about licensing their IP should they find something of interest that would have been a more just outcome and likely would have yielded more proposals. As one ancient Israeli philosopher once said "You must not muzzle an ox to keep it from eating as it treads out the grain.". This means that the workers (in this case, the researchers) should not be muzzled from being rewarded for their work. Unlike almost any other profession, academics have their ideas as their most important asset, and in order for America to remain the global leader in innovation, academic researchers must feel their ideas are protected and even redeemable for reward in the form of licensing, when the ideas are of economic value. We should remember that the Bayh-Dole act was born to remedy a situation where the innovation ecosystem had stalled because idea and IP generation was de-incentivized.

The merits of university-federal lab/agency partnerships. Government laboratories serve many noble purposes for our nation from an academician's viewpoint. First, they provide our government with the research capacity in terms of personnel and equipment infrastructure to tackle the nation's pressing scientific problems, both generally and on-demand. Second, they provide a rich research ecosystem of researchers who care about the science of discovery divorced from the pressures of generating quarterly profits. And third, they provide collaborative resources in terms of intellectual capital, equipment, and mentorship for young researchers.

Disclosing my own background and educational support as a student. Let me start by disclosing that NASA funded my entire education from my freshman year of college at a Historical Black College and University (HBCU), Tennessee State University, until I completed my doctorate at a small, private, technical university in upstate New York, called Rensselaer Polytechnic Institute (RPI). Specifically, it was NASA Glenn Research Center who funded my education through a variety of scholarship and fellowship programs. Over those years, I personally engaged with no less than 50-75 PhD holders and PhD students. They raised the bar for me in terms of what serious scholars were like, what they got excited about, what they did to elevate themselves above the average group of scientists and engineers, and finally, why they were so excited about mastering the literature and writing publications. NASA Glenn, a federal research center, was invaluable to my development as a scholar and researcher.

Interactions with federal center/labs as a professor and researcher. As a Professor and researcher, my lab interacted with two government centers/labs in a major way: (1) NASA Glenn Research Center; and (2) the National Energy Technology Laboratory (NETL).

First, NASA Glenn supported my lab by collaborating with us on a project to test a super-elastic material known as Nitinol. They brought the project to us, and it resulted in several papers, although no direct funding. We were able to leverage their materials processing facilities to fabricate test structures and we engaged with many of their prominent material scientists who went on to write reference letters for my students and convinced them get active in professional technical societies in our field. They also sponsored one of my former students with a NASA GSRP Fellowship, which gave him PhD direct funding and access to resources such as an air erosion tester, high-temperature tribometers (which we do not have access to), and profilometers—all equipment used to test the effects of space dust and contamination on lunar surfaces. Again, the mentorship my former student received was excellent; government labs really do reinforce and accelerate a graduate student's path to research maturity.

Secondly, my major lab interaction has been with the Department of Energy (DOE) NETL lab in Pittsburgh, PA and Morgantown, West Virginia. NETL introduced my lab to the oil and gas drilling problem. They had just purchased a million dollar ultra-deep, high-temperature, high pressure drilling simulator which we could not get access to anywhere else. We were tasked with modeling the phenomena it would measure. They also provided us with wonderful rock samples they machined in-house to allow us to test on our own equipment. As a result, we became reasonably well known in the drilling services industry for our highly complex computer models of the drilling process. NETL also granted us time on their supercomputers and supported postdoctorates in work there NETL. Again, the mentorship was an excellent benefit for my students. I also participated in work involving granular media with the Multiphase Flow group at NETL. I can honestly say that just like many of my other colleagues who work with government labs, their support of our research in indirect and direct ways was pivotal in me progressing up the research path and emerging into a leader in my field, which of course helps with personal things such as promotion up the tenure track. Because of their regional location, they were interested in extending the facilities and resources to universities that were in their region working on research they deemed aligned with their missions. We are grateful to these two labs.

Other important but less major interactions with federal centers/labs. There are also numerous other labs within NASA (NASA JPL and NASA Johnson), and the DOE (e.g., the National Renewable Energy Lab, Argonne, Sandia, Lawrence Livermore, etc.) who have hosted my research group for visits, hired my students, engaged us on data, and written support letters for us to obtain funding from the NSF. We are in conversations with some of these entities because of their vast research capacity and state-of-art equipment (e.g., an open architecture 3D printer for testing new material processes for fusion 3D printers). One of our collaborators performed amazing imaging of single powder micro-particles at Argonne National Lab's Synchrotron X-ray microtomography, where the pores in a single powder particle were visible. Such an instrument is cost prohibitive for most universities. Further, some universities have centers, such as my own university's Energy and Environment Initiative (EEi) that works to connect its own academic researchers with those from federal labs in order to streamline the access to state-of-art infrastructure and collaborations with world-class researchers.

Lastly, some federal labs have even provided guidance to start-up companies such as my own, InnovAlgae. They often help SBIR companies outline the best paths to technology validation, including connecting them to industrial partners that could benefit commercialization efforts.

The merits of university-company partnerships.

Universities use money for basic research to create knowledge, but companies use the knowledge from that same basic research in order to generate money.

When this cycle is done well, companies who understand the academic research enterprise are pivotal components in the cycle of training, mentoring, graduating, and employing the nation's best and brightest students. However, there are more often than not, a rising tide of companies who struggle to see their crucial role in this cycle. A company that once told me a similar version of the above quote (which I have modified to serve the theme of this testimony) would often want to collaborate with my lab on NSF Grant Opportunities for Academic Liaison with Industry (GOALI) proposals, which are NSF mechanisms to allow research proposals to be a little more applied, and aligned with industry. The partnering company does not have to offer anything more than in-kind contributions which can take on almost any form. Many companies therefore do not spend their resources on the research they seemed to covet for academic labs to generate. Since the NSF engineering program funding rates are around 10-15 percent, it means that my team and my collaborating company would have to submit 7 to 10 proposals over many years (some have a single annual submission window, while others have just two) before a proposal is funded. Of course, most researchers eventually move on and the collaborative opportunity, which are often brilliant ideas, flounders. It would be a game changer if the company were incentivized to invest seed money into the basic research of the project they like.

Yet, when some of these companies have job openings, they often bombard us with emails to get our graduating students. I used to wonder why they expected me to have any students at all when they broke the cycle (i.e., universities use money to create knowledge; companies use the knowledge to generate money). This is not every company of course, but such interaction is common and this committee can help us both the university and company's efforts.

I think the Congress can provide some win-win help here by looking at what some other governments do well, at least in this one area. For example[7], Brazil's government requires oil & gas companies to spend ½ of 1% of their gross revenues from Brazilian oil field production on Brazilian universities.

Perhaps more aggressive tax incentives could be offered to companies who invest in American research universities' basic science research. They can give these universities some percent of their gross revenues and receive increasing tax incentives when they go above some threshold. These companies would

---

[7] Source: Andrews, Phil, Jim Playfoot, and Simon Augustus. *Education and Training for the Oil and Gas Industry: The Evolution of Four Energy Nations: Mexico, Nigeria, Brazil, and Iraq.* Elsevier, pp.70, 2015.

benefit from having a stronger American workforce, richer basic research results from which to learn, and the country would benefit by having its capacity for basic research increased.

A final recommendation for supporting basic research. If congress were to inject new funds into the NSF to increase the number of NSF graduate student research fellowships by a factor of two from 2000 to 4000 per year (the total annual number of applicants is typically 14,000) which is likely to increase the NSF fellowship investment from $333M to $666M annually, this would inherently: (1) fund basic research since NSF fellows have academic freedom to work with faculty to choose any project, which will likely be more basic science; (2) strengthen the American science, technology, engineering, and math (STEM) PhD pipeline since the high stipend levels have been shown to motivate more of the nation's 'best and brightest' students to stay in school and pursue doctorates; (3) this would loosen up the tension of the low funding rates for basic science and it will also allow principal investigators to support a diverse range of 'crème of the crop' students from other nations in addition to our own students not supported by the NSF fellowship.

Dr. Fred Higgs
John and Ann Doerr Professor of Mechanical Engineering
Rice University

Dr. Higgs received a B.S. in Mechanical Engineering (1995) and an M.S. in Mechanical Engineering (1997) and Ph.D. in Mechanical Engineering (2001) from Rensselaer Polytechnic Institute in Troy, N.Y. From 2001-2003, he was a Postdoctoral Research Fellow at Georgia Institute of Technology. Dr. Higgs was a faculty member in Mechanical Engineering at Carnegie Mellon University from 2003-2016. He joined the faculty at Rice in 2016. Dr. Higgs is also the Faculty Director of the Rice Center for Engineering Leadership.

Dr. Higgs has received a National Science Foundation CAREER Young Investigator Award in 2007 and in 2013 he received the Benjamin Teare Award, the Carnegie Mellon College of Engineering top teaching award. He is a member of the American Society of Mechanical Engineers and became a Fellow in 2016. In 2010 he was presented with the Burt L. Newkirk Award for his notable contributions to the field of tribology. He is also a member of the Society of Tribology and Lubrication Engineers, the National Society of Black Engineers, the Materials Research Society, and the Institute of Electrical and Electronics Engineers.

Chairman WEBER. Thank you, Dr. Higgs. The Chair now recognizes himself for five minutes.

Dr. Tirrell, your written testimony touches on an important quality of the national labs: the core capabilities and user facilities that allow a single researcher to use a number of tools at a single lab to make a scientific discovery. What steps could the department take to better streamline access for those researchers across the lab complex?

Dr. TIRRELL. Thank you. Well, I think we at Argonne have several major research facilities, as you alluded to, in x-ray scattering and in computation, also the Center for Nanoscale Materials. What is really important is the staff scientists that staff those facilities because users often come with an idea of how to—of what they want to do but not necessarily how to do it with our facilities. So we need experts on site really to try to make the time that they have on the instrument which sometimes is 24 hours, and it can go from, you know, straight 24 hours the most effective. So you can get there, get in and get out with the results that you need. But that requires dedicated and really knowledgeable staff. I think that might be the principal thing that I would suggest.

Chairman WEBER. Okay. So we have to schedule facility upgrades like the Advanced Photon Source. How important are those upgrades for providing the tools needed for the materials research community?

Dr. TIRRELL. There's nothing more important than the upgrade to the Advanced Photon Source, both for Argonne or for the x-ray scattering community in the United States. It's really the only state-of-the-art hard x-ray, meaning high-energy x-ray, facility in the United States that can do certain things, and many other countries are investing but what we need is a facility that U.S. scientists can access most effectively.

Chairman WEBER. Do you know where we are on that scheduled timeline?

Dr. TIRRELL. Well, a lot of that depends upon the rate of funding. Right now, we would have, I believe—I hope I don't misstate it but roughly dark time, meaning that the equipment would be installed—the new equipment would be installed in fiscal 2020 and then come up for operation later in fiscal 2023. If the funding that's proposed now is maintained, this could be delayed by a year or more, so that's the kind of time scale that we're talking about, and the implications of the funding profile.

Chairman WEBER. Okay. Thank you.

Dr. Schwartz, this question is for you. In your prepared testimony, you talked about a project currently underway at Ames Lab using caloric materials to improve the efficiency of heating and air conditioning and refrigeration. Could you describe what these materials are and how they may change the industry for us?

Dr. SCHWARTZ. Okay. I'll try and make this relatively quick and simple. Caloric materials are a type of material that when you apply a field, a magnetic field, an electric field or a stress field, there is an internal change in the structure that creates a significant temperature change. So now you can imagine having a closed system where you have a warm fluid coming in, you have your magnetocaloric material, for example. You apply a field. It changes the temperature. It cools that fluid coming by and you have a refrigeration system. It won't use greenhouse gases. It will be environmentally friendly, and if constructed with affordable, Earth-abundant, easily manufacturable materials, it could potentially transform the refrigeration and air conditioning industry.

Chairman WEBER. How do you move that fluid? You know, you use a compressor to remove refrigerant that change the state twice in the typical refrigeration system, so how are you moving that fluid?

Dr. SCHWARTZ. I think that would be the same way. You'd have a pumping system that would either bring in the air or the liquid over top of the caloric material.

Chairman WEBER. So instead of a compressor that compresses refrigerant from a loosely packed gas into tightly packed liquid and diffuse it through a metering device and it sprays out and has a temperature drop, pressure drop and it picks up heat there, is there a metering device? I don't know how much work you all have done on this. This is fascinating to me. In fact, what time is it? We may be here for a day or two. So is there a metering device in this system? How do you get this corresponding temperature and pressure drop in that system?

Dr. SCHWARTZ. So our research is focused primarily on discovering new materials in order to enable this technology to go forward. And I'd like to point out that the first material of this type was really invented at Ames Laboratory about 20 years ago. Research was funded through basic energy sciences as it had been for some time before that. After the discovery of this material, it wasn't long before industry said hey, we got this, were going to make something good out of it. As a result, basic energy science has said okay, industry's got it, that's out of our realm, we're not going to fund that anymore. Well, 20 years has gone by and industry has not been able to pick up that technology because of the inability to do the basic materials research and enable that amazing new material technology to be implemented into something as impactful as revolutionizing the air conditioning and refrigeration industry.

Chairman WEBER. Well, the fact that it's 20 to 25 percent of energy consumption, as was pointed out, you know, is a pretty astounding figure, and we could go on for a long time, but I'm going to go ahead and—who am I yielding to? Marc has left, so I guess, Daniel, you're up next. The Chair recognizes you for five minutes.

Mr. LIPINSKI. Thank you, Mr. Chairman.

I want to start out with Dr. Tirrell. Argonne is home to the Energy Innovation Hub called the Joint Center for Energy Storage Research, more commonly known as JCESR. It had great success since 2012, but the Trump budget proposes to eliminate it.

So I wanted to ask you what would the consequences of eliminating this Energy Innovation Hub be, and would private sector be likely to pick up this work?

Dr. TIRRELL. Thank you. Well I think it would—the cutting of JCESR would leave a lot of very promising research results on the table without further development. There have been industry involved in JCESR, Johnson Controls, for example, which is the largest battery manufacturer. But again, it could be a situation such as Dr. Schwartz described where while the technology is promising, it's not really sufficiently developed that a company is able to take it over.

Having said that, it's conceivable that there would be ways for the Office of Science to continue its investment in energy storage research at Argonne and elsewhere in the DOE complex. So we don't view it as a great thing that JCESR may be coming to an end, but I think that it has already produced a wealth of results that can be followed up on if additional investments are made.

Mr. LIPINSKI. Well, we have to—I think we need to fight here in Congress to make sure that we don't defund these Energy Innovation Hubs for what they are doing, where they have gotten so far in the research and development.

But I think that really leads me to my second question for Dr. Schwartz first and then Dr. Tirrell. There's this false boundary that's being claimed between basic research and applied research, and saying well the federal government—some will say the federal government should only be involved in basic research and not applied research. I don't think that there really is a neat divide here, and Dr. Schwartz, you mentioned in your testimony your concerns about so much that would not be done if the federal government just got out of the development part of the R&D research and development sphere.

Can you tell me why that is and why the government needs to be involved in the development?

Dr. SCHWARTZ. There is a common view that research from grand challenge and basic science is just a continuum, and that once you start on that path of understanding, that that's going to take you to the logical conclusion that could ultimately be commercialized.

In my experience, I have never seen anything like that. We make progress in one area that opens up new doors. We might explore that path and then have to come back, so there's the pipeline model of technology development that only applicable a few percent of the time. There's another model that shows more of a feedback loop where instead of having just one valley of death in the commercialization of a product, you actually have two. One is taking a look at the feasibility of the product or of the material, and of course, the second is the late stage, being able to scale up and commercialize it. It is not a linear path between discovery and implementation. Sometimes, like the case of the caloric materials that I just talked about, it looked like it was promising but no one had done the full development of the materials to make that feasible as a commercially available material.

So the feedback loop happened. The material was discovered. Industry thought they were going to pick it up, were not able to or chose not to invest as much as they needed to to get that product available, and then now energy efficiency and renewable energy, through one of its recent energy materials networks, has picked up that research again to do the foundational science required to create the new materials that will enable this technology.

Thank you for your question.

Mr. LIPINSKI. Dr. Tirrell, anything briefly you want to add?

Dr. TIRRELL. Yes. Certainly I agree with the premise of your question and some of the things that Dr. Schwartz said. I used the terms iterative and cyclic and non-linear a couple of times in my own testimony.

One thing I point to is the Office of Science Basic Research Needs workshops—there's brochures about them out in the hallway—where the Office of Science tries to define important basic research in quantum computing, in water, in synthesis, based on what's needed to carry these things forward into practical technology. So I think we all recognize this interplay between basic and applied research, even just as an intellectual thing in addition to its practical implications.

Mr. LIPINSKI. Thank you. I yield back.

Chairman WEBER. Thank you, Dan, I appreciate that. And I do want to add, though, that we did—House did pass H.R. 589, the Energy Innovation Program, where all of those hubs are actually authorized, and unfortunately it's sitting over in the Senate and we just hope the Senate has enough energy to get something done. Did I say that out loud?

I now recognize Barbara Comstock for five minutes.

Mrs. COMSTOCK. I don't have a Madonna quote. I'm speechless.

Dr. Locascio, how does a prize competition like the Head Health Challenge promote the development of new materials, and what did NIST learn from participating in this prize challenge?

Dr. LOCASCIO. Thank you for the opportunity to speak about that. The prize challenge really is just one tool in our tool kit to leverage federal dollars against private sector dollars, and I think it's an extraordinarily effective way to do that, and also to pull in people into an important national problem that may not have been aware of or how to get involved. And so I think the prize challenge that we conducted with Head Health, it's a partnership between NFL and NIST and GE and Under Armour, has been very successful in attracting new people into the problem associated with public safety, and in particular, protective gear.

For instance, we had people competing in the prize challenge who presented new materials that were additively manufactured or prepared in the laboratory that were responsive materials or new types of materials made with new processes that had never thought before about using them and harnessing that activity for protective gear.

So I think one advantage is really being able to attract new people to these new national problems, and our role there is really to help conduct an unbiased and fair competition, and we were able to leverage testing equipment that we already had developed for the purpose of testing headgear for war fighters, and used that to conduct these tests. And in the same time, push forward our capabilities even further into new realms to test these types of materials.

Mrs. COMSTOCK. And how—and maybe some of the others can add to this, too. How can you—how can we develop more of those partnerships like that, because I think the synergy is there. The relationships really cross over so many different disciplines. It's really exciting. You're getting a lot of different partners who have a lot of different interests in this. So how can we build on that model and find some other areas, and what are some other examples that we might pursue in this area?

Dr. LOCASCIO. So I'll perhaps start and let others chime in, but we've learned a lot from NASA, who was conducting prize challenges about how to leverage the external community and attract them into these types of prizes.

This was the first one that NIST had conducted and the first one the Department of Commerce had conducted, and we've gotten so much out of it that we currently have several others in the pipeline, current prize challenges that are being awarded soon.

Dr. HIGGS. So I've been actually—I've been on the side that actually is the competitors for these different challenges, and I do admit that when these challenges come out, my students and I, you know, all want to be competitors in some sense, maybe athletes or something, and we see that as our opportunity as researchers to compete, and we always think we're going to win, of course. But these, you know, competitions have a really good basis for being able to generate ideas and things, and we love it when the government labs are involved with doing these as well.

Certainly, we would just caution that, you know, sometimes when industry is involved with these competitions, they—I've been with several colleagues and you write a proposal and at the end of it, it will say any idea that you submit, we can actually take. You're giving up your rights to that particular situation. So I would just say make sure there's oversight, certainly, when there's industry there, because we don't want an awesome idea to be used as a way to backdoor and take IP from universities that could generate revenues to do other important things with basic science.

So love the competitions, but we'll just say some oversight when the industry is involved, making sure that IP is not given up in the wrong way.

Mrs. COMSTOCK. Thank you. Thank you, and I yield back, Mr. Chairman.

Chairman WEBER. Thank the gentlelady, and Mr. Veasey, you are now recognized.

Mr. VEASEY. Thank you, Mr. Chairman. This question is for Dr. Schwartz.

Dr. Schwartz, in the fiscal year 2018 budget proposal, Ames Laboratory did not fare well. If this budget were enacted, it looks like your capabilities and scientific workforce would be decimated. I was wondering if you could lay out the consequences of this budget proposal for Ames Laboratory, and if enacted, do you have an estimate for how this would impact your workforce?

Dr. SCHWARTZ. Thank you for your question. The proposed budget that I've seen for Ames Laboratory proposes a 58 percent decrease in the budget between the fiscal year 2018 request and the fiscal year 2016 enacted. Clearly, a 58 percent decrease in the overall budget is going to have an impact on our staff, and it is also going to have an impact on our ability to meet our mission to create materials and energy solutions.

Mr. VEASEY. How would this budget proposal impact materials research at Ames and, you know, largely how would it affect it in the U.S. as well?

Dr. SCHWARTZ. The work that is going on at Ames Laboratory, other national laboratories, universities, NIST is successful because of the long-term sustained federal investment. Science is something that progresses continuously, sometimes quickly. More often, not so quickly. Interruptions to that flow of science would be significant. Decreases in scientific staffs at the national laboratories certainly slows down projects, if not stops them. It makes it more difficult to pick it up.

In addition, the potential decrease in funding in the materials areas sends a message to high school students, college students, early career researchers at universities, and assistant professors, and I'm not sure that's a message that we want to send. Materials research has been demonstrated to provide economic value, energy security value, national security value. I would like to see that progress continue at a rapid pace.

Mr. VEASEY. This is sort of regarding the first question I'd asked you about your workforce. Could you be specific about exactly how many people would be laid off or what numbers your workforce would be reduced with these budget cuts?

Dr. SCHWARTZ. We have done an estimate based on that 58 percent decrease from the '18 proposed to the '16 enacted budget, and assuming that we do not use funds that are carried over from existing what we have now, we're looking at a decrease in the overall staff approaching 40 percent.

Mr. VEASEY. Thank you very much.

This message is for Dr. Tirrell. I know that the drastic cuts proposed to the budget would have major consequences for our Argonne National Laboratory. I was wondering if you could also walk us through the impacts that this budget proposal would have on the capabilities and workforce of Argonne if it were enacted.

Dr. TIRRELL. Yes. Obviously if those cuts are enacted, the capabilities in the spirit of Chairman Weber's question about how we could staff user facilities may be affected. Cuts will affect our capabilities and workforce. Partly as a measure to protect morale, we haven't made public statements of, you know, exact estimates because we don't know for sure what's going to happen. There was a, you know, a business newspaper in Chicago that suggested that the cuts would be something like 700 combined across Argonne and Fermi lab, but that's an independent estimate that we are not part of. But clearly, it will impact our capabilities and workforce.

And you know, a thing that's important to recognize, and it's true of national labs, university labs, and industrial labs, they're much easier to tear down than they are to build back up after that, so it's an important step to think about.

Mr. VEASEY. And also I wanted to just ask you specifically about your portfolio of material research at Argonne. Can you just very quickly say how that would be impacted?

Dr. TIRRELL. Well as I mentioned in my own testimony, we do span in several areas such as energy storage from electro chemistry to battery prototypes. As I understand, the budget proposal would be hit more heavily on the applied end of that, so how well we could get things to the point that the commercial implementation, I think, would be the place where the pressure would be applied by these budget cuts.

Mr. VEASEY. Thank you, Dr. Tirrell.

Mr. Chairman, I yield back my balance of the time.

Chairman WEBER. All right, thank you, Mr. Veasey. We now recognize Mr. Dunn for five minutes.

Mr. DUNN. Thank you, Chairman Weber. Good morning to the panel. My name is Neal Dunn. I'm a medical doctor recently turned Congressman from Florida, so the chance to listen to so many great scientists is a real pleasure for me. This is my only dose of science I get, really, up here in Washington, so thank you very much.

In our district, we have Florida State University, one of the preeminent research universities in the country. We have a new material science and engineering program there that is rather large, but perhaps most famously includes the National High Field Magnetic Lab. I suspect maybe you collaborated with them from time to time, and I'd like you to keep that in mind as I make the comments and ask my questions.

I'd like to start with Dr. Higgs. First, Dr. Higgs, I want to encourage you to think of your sojourn in Texas as temporary. I know that——

Chairman WEBER. The gentleman's time has expired.

Mr. DUNN. The sugar white sands are calling to you even as we speak.

You actually said something very important, especially in this time of compressed budgets, and it was about the university IP. So historically, I think universities, as you say, they turn money into knowledge and they may spend $100 million in a year and then on royalties they'll get $1 million back. Well that's a very poor return on investment. I think we all recognize that. Now there's many universities, I'm sure some of the leading ones that you deal with have adopted newer techniques, but it's important, I think, that we push this out into the labs as well, these partnerships, because you're right. Your faculty and your post-grad students take with them IP into the private sector, and they try to monetize that. And I think we can keep them in the faculty, keep them in the labs—your labs if we share the IP, the ownership of the IP in a more intelligent fashion. I think you're doing that. Am I right? Answer that, Dr. Higgs. It sounded like you have some familiarity with how to parcel out the IP—the rights of the IP so that you kept the talent and the ideas still got to market.

Dr. HIGGS. Right, good question.

So first of all, I want to say I'm originally from Tallahassee, Florida, so your district, and I did participate in pre-college engineering programs that motivated me to pursue a Ph.D. in mechanical engineering. It was at the Florida State University and Florida A&M University, minority introduction to engineering.

Mr. DUNN. Come on back. The water's fine.

Dr. HIGGS. Right. It was at this program where I had the sophistication to realize that a terminal degree was the way to go, so I thank your district for supporting young dreamers like me.

Certainly we, you know, we have a responsibility to our employer, the university, that whenever we generate an idea that the idea belongs to them because of the Bair-Dole Act, and—but we are really most interested in working in basic science. But we're in a capitalistic society, so these things have to be funded. And you're right, some companies fund us and we do the research. The companies will ultimately get our students there. The IP that's in the university, the whole goal of it is to actually get into the market to help——

Mr. DUNN. Chairman Weber is going to cut us off quickly, so I'm going to say that I encourage all of you to think of it as public-private partnerships and really help—that helps monetize your lab—monetize the ideas, but also keep the people in your lab where you want them.

Dr. Schwartz, regarding your caloric material on refrigeration. You have a cooling source we have in Tallahassee a company that manufactures a frictionless bearing. It's a magnetic bearing, no lubrication, and they turn in 20, 25 percent savings on industrial HVAC units. I think, you know, we've got a marriage here. I'm playing matchmaker. So I think you've got—you put those two things together. Somebody removed the fluid or the air in Tallahassee. In fact, my staff will no doubt share with you the name of that company so that you can work with them.

In the 30 seconds remaining to me, Dr. Locascio, how do you define success when you're looking at grant applications? What makes you find a great grant?

Dr. LOCASCIO. So we go through a peer review process for all of our grants. It's a very well-structured process, and it's pretty common across——

Mr. DUNN. There's no hook right now? You're in a low monetary budget kind of finance. What do you do? What are you looking for?

Dr. LOCASCIO. Oh, how are we pursuing grants? Are we going to continue to pursue grants?

Mr. DUNN. Well, our time expired, but—and I've already tested the Chairman's patience, so——

Chairman WEBER. No, go ahead. I'm interested in her answer.

Dr. LOCASCIO. So we will continue to put out grants to universities. Obviously, we've had very hard decisions to make as well with regard to the budget, but one of the things that we've thought about is really protecting the future. And protecting the future means also protecting our abilities to do the greatest advances in measurement science that you can possibly do. And that honestly requires the universities. We have to collaborate with the smartest minds in the United States and pair them with the smartest minds in the federal government, and we do that to great benefit. So we'll continue to put out grants.

Mr. DUNN. Thank you very much. Thank you, Mr. Chairman.

Chairman WEBER. You bet.

The Chair now recognizes Mr. Foster for five minutes.

Mr. FOSTER. Thank you, Mr. Chairman, and before I return to this—today's symposium theme on magnetocaloric refrigeration, I would like to make just a few comments about, you know, the elephant in the room here which is the proposed draconian budget cuts to the entire laboratory system.

Earlier this month I was joined by 55 of my colleagues in sending a letter to the heads of all seven science agencies, asking about the impact of our Republican President's budget request on jobs, not only at our national laboratories, but at our universities that rely on federal funding to train the next generation of scientists. We have yet to receive any response to this, and I think that, you know, despite the risks that have been mentioned to the morale of everyone involved, I think it's important that we look this dragon in the eye and make sure that all Members of Congress who claim that they support science speak up at times when science funding is at this kind of threat.

So without objection, I ask unanimous consent to submit this letter to the record.

Chairman WEBER. So ordered.

[The information appears in Appendix I]

Mr. FOSTER. Now to get back to the fun stuff.

Materials science, you know, like many other disciplines, has benefited very greatly from R&D funding. And so actually, I'll return to Dr. Schwartz for a second, despite the fact that you do not—you're not part of my constituency as Dr. Tirrell is.

You know, you mentioned a 40 percent—40 percent is a rough estimate for the layoffs. When that sort of thing happens to a technical staff, if future administration or future Congress decided to just restore that, can you just throw a switch and immediately regrow the technical expertise that's been lost, or is it more complicated than that?

Dr. TIRRELL. I think it would be much more complicated than that. Scientists who either choose or are forced to leave their jobs will look for others. I don't believe that private industry is going to be able to pick up all the researchers that would become available through this budget. They would then search to change their fields. We have many researchers at Ames Laboratory and across the National Laboratories system who have come from foreign countries. There would be a significant risk that many of those scientists would return to their home countries. They would take their education, their experience, all of the investment that we have placed in them, free of charge, back to their country. Right now, we are trying to extend our global leadership in materials research. I think slowing down that progress and then restarting it later would be quite a challenge.

Mr. FOSTER. And is—maybe someone else on the panel could comment on the effect that that would have on the morale of younger students coming into the field or post-docs coming into the field when they see, you know, massive layoffs in their often very focused field of expertise? Dr. Higgs?

Dr. HIGGS. Well that's what my notes were actually saying. That's a very perceptive question. I mean, if you think of a lab like, say, Ames or Argonne in particular, Argonne has some—in my area, they have some very prominent tribologists, and essentially what happens is if the tribologist is just a little known, once they are removed and someone says hey, you know, this particular scientist no longer has a job, then the entire community goes what does this mean for tribology? Should we all try to head for Silicon Valley? Should we all do something with the right now implication as opposed to a long-term implication, which is what research science has? And then the younger students, we have to give a speech to encourage them to stay the course, but yet we're uncertain as well.

So definitely, even though we're in a university environment, whenever there's a cut to a prominent area or prominent scientist, once they're removed from the equation, there's a lot of questions that we have to answer as academicians and the students are asking about that. Excitement and morale definitely takes a hit.

Mr. FOSTER. Thank you. I think that's a very important thing for Congress to understand, that this is not at all like, say, starting and stopping a highway construction project. You can't just throw the switch and recover the damage that was done.

Sorry. Now let's see. I have a little bit of time left, so I'm going to return to magnetocaloric refrigeration. It's my understanding that the fluid—the working fluid that's here is largely just a heat transfer fluid. There's no phase change involved, and the compressors involved are a small fraction of the total power to do—perform the refrigeration. Is that a correct understanding, or is it more subtle than that?

Dr. SCHWARTZ. There are lots of details on how you would implement a solid state magnetic or electric or stress-induced cooling system. Our focus right now is the very, very early stages. Can we develop the materials in order to—that demonstrate, that have those large temperature change. At that point, we will turn it over to our mechanical engineering friends who will then design the system, optimize the fluid flow, heat transfer, and others. Right now, our focus is really on creating those materials that will enable the transformation in air conditioning and refrigeration.

Mr. FOSTER. And the rest of the problem, just the getting the heat transfer fluid across the plates or whatever they are, is a—it's closer to being a solved problem and an engineering optimization. The magic is the material that you have to make work and at high efficiency, high lifetimes, all the challenges?

Dr. SCHWARTZ. That is my understanding, yes.

Mr. FOSTER. Okay, good luck. I really look forward to having refrigerators that don't rattle in the middle of the night.

Dr. SCHWARTZ. And last much longer. Thank you for your question.

Chairman WEBER. If you'll quit getting snacks out of the doors at midnight, then you won't hear that rattling.

The Chair now recognizes Mr. Marshall for five minutes.

Mr. MARSHALL. Thank you very much, Mr. Chairman.

Both Dr. Schwartz and Dr. Higgs mentioned 3D printers. Chairman Lamar Smith and I recently got to go to Wichita State University and see the largest 3D printer in the world, about 1/3 the size of this room, and we'd love to invite you all to come see what they're doing there on their innovation center, always believe it's opportunities to promote each other and work together.

I should ask Dr. Schwartz and Dr. Higgs both what they see—what's next for 3D printers, specifically, you know, what's going to be a game changer? What type of more viable mass do we need? What do you see next for 3D printers? Dr. Higgs, do you want to go first?

Dr. HIGGS. Very good. Thank you for the question.

I mean, definitely if you think about it, when you look at Star Trek you don't see really big engineering manufacturing labs. You just see something very small, and they ask for the product to be developed. And so that's kind of where attitude would have to hit. So you would want essentially to be able to additively manufacture anything, and that means that you have to be able to work with multiple materials. Right now you see a beautiful 3D printer, but it only prints a limited amount of, say, materials that are there. So this big one that you talked about is probably a metal printer, and if it is, it's a limited set of materials. But if you want to print something that's, say, biocompatible, then you may not be able to use steel or gold or something like that, and so you need to be able to change out the different materials. If you want them additive, you can build them part by part. You want the mechanical properties to change as you want them to, then that means you have to have a functionally graded material, which means that it may start one mechanical property at one end, and be another at the other end. Right now that can't be done, and so there's some important material science questions that have to be answered.

But the point is that you want to print anything as you want as it could occur. Additive can do that in principle, but the basic science questions have to be answered to unveil that to the society.

Mr. MARSHALL. Dr. Schwartz, anything to add?

Dr. SCHWARTZ. That's an excellent question, and to me, the key to successful deployment of additive manufacturing in this growing industry in the U.S. is all about understanding the materials properties. Researchers have been trying to understand details of steel, aluminum, titanium alloys for decades, if not centuries, and they still don't have full understanding. Now we want to make additively manufactured parts out of the same materials, but the process is so much different. The composition will change as you melt and re-melt as you make the powders.

Right now, I believe the key is getting a fuller fundamental understanding of—starting at the very beginning, developing the metal powders. Without the metal powders, none of the metal additive manufacturing happens. Those powders have to be pure. They have to be spherical. They have to flow nicely. They have to have the right surface conditions, and all of this is based—we need that basic research understanding to get there. No one has ever looked at laser melting of particles in great detail. This is a brand new field. Ames Laboratory is working with SLAC and Lawrence Livermore National Laboratory and using one of the national user facilities, Stanford Synchrotron Radiation Laboratory, in order to understand that the early stage materials melting and resolidification and development of that most important internal structure that's going to control the properties. It's a very exciting time.

Mr. MARSHALL. It is. One of the exciting things I saw was they build you to take away from the product that it's printing and telling the machine to maximize it, so they were doing wing replicas and trying to have a stronger wing for airplanes, for jets, but yet lighter, and to see that technology come forward. So it is very exciting as a physician to see what they're doing in joints, to think that instead of having your choice of hip joints as small, medium, or large, you can actually make one that fits your joint is exciting.

Last question for Dr. Higgs. I see that you won the NSF Career Award, so congratulations. Professors at Kansas State University, which is the champion of the Texas Football League this past year, having defeated——

Chairman WEBER. This gentleman's time is also expired.

Mr. MARSHALL. —Texas A&M, TCU, and Texas Tech. Anyway, professors at Kansas State University, Wichita State University, and University of Kansas have all won that recently. Tell us a little bit about that and what you're doing with that foundation grant, please.

Dr. HIGGS. Very good. So I had an NSF Career Award. It's supposedly given to the nation's youngest—best young researchers. And I do want to say that the research from that, which was actually to develop slurry technology, was about five years after that grant. It became an NSF SBIR company, InnovAlgae, that I now have. And so it's making it back up to the marketplace because of the basic science research that's now translated into a small company. Thank you.

Chairman WEBER. The gentleman yields back. Ms. Bonamici is recognized for five minutes.

Ms. BONAMICI. Thank you very much, Mr. Chairman, and thank you to all of the witnesses for being here today.

I want to start by aligning myself with Dr. Foster's comments about education and the message that these budgets cuts send, both to students who are contemplating graduate school or students who are in undergraduate trying to decide their career path. I just came from the Education and Workforce Committee on which I serve, and have as a priority wanted to make sure that we are educating people here in the United States for the jobs of tomorrow. I am very concerned about the sort of shift in the message that we're sending.

There was a time when federal funding for research and development was growing and graduate students were optimistic about careers in research. We need to get back to that message to our students and our potential new scientists across the country, and I'm very concerned about that, and our leadership. And I just point out as one recent—very recent example that disappointing decision to exit the Paris Climate Accord, and then immediately France started recruiting our scientists. We need to have U.S. leadership here and maintain that leadership, and I'm, again, very, very concerned and share the concerns of others about what these budget cuts— what the message is to students and to the rest of the world.

I'm—Congress really needs to think holistically and long-term about supporting the sciences. I'm concerned about multi-year projects which Mr. Foster mentioned, and I've heard from scientists in Oregon who are very concerned about the lasting effects of these cuts to their research, to the country, to our leadership, and the global community.

Dr. Higgs, could you speak briefly about the concerns of your students when they're considering continuing careers in research? How do you advise them about their future careers in light of these uncertainties and proposed budget cuts? And I do want to save time for another question.

Dr. HIGGS. Very good question. So we definitely are always trying to aim them at going to academia, a government lab, or an industry. We would really like to work on basic science, because we know fundamentally that will translate into anything there, but you become more constrained as cuts come. Cuts usually—government cuts usually mean that basic science is out, so then we have to work on some specific problem, and so then we become people who are out looking for funding all of the time, rather than educating, because we have these young, bright minds we really want to go through and get a Ph.D.

Ms. BONAMICI. Absolutely.

Dr. HIGGS. So we look at mentoring them. Government labs we work with, they also mentor our students as well.

Ms. BONAMICI. Right. We want them to get their Ph.D. and stay here.

So in the President's—this is Dr. Schwartz. In the President's budget proposal, the Office of Energy Efficiency and Renewable Energy would receive a 70 percent cut. The Renewable Energy and Sustainable Transportation portfolio, 70 percent reduction. Energy efficiency, 80 percent cut. This is concerning. Clean energy jobs are an important driver of our economy and the research helps advance these industries. In fact, the Bureau of Labor Statistics found that wind turbine service technicians—it's one of the fastest growing occupations. Many of those jobs are in rural areas. How would these massive cuts to EERE affect materials research at your labs and in the clean energy industry, and how would they affect the growing—rapidly growing clean energy job sector?

Dr. SCHWARTZ. Specifically for Ames Laboratory, we have really four main projects that are funded through Energy Efficiency and Renewable Energy. The Critical Materials Institute, one of the four energy innovation hubs, a very important scientific endeavor, early stage basic research that is supplying critical options for the United States moving forward with regard to rare earths and other critical materials.

Just last week the only mine in the United States that was producing rare earth materials was sold. We now have no capability to mine rare earths. That's a big concern for me in terms of economics and in terms of national security.

Another big project, the caloric materials consortium that we've spent a little time talking about today, that is also funded by EERE. The powder synthesis work that we are doing, trying to create optimized metallic powders to enable the 3D printing industry, that is also funded by EERE. All of those are in jeopardy if this budget goes through.

Ms. BONAMICI. And in my remaining time, could you, Dr. Schwartz, address—the President's budget declared some research at an early stage worthy of federal support, and other activities as later stage research that should be immediately eliminated, given that the private sector is supposedly better equipped to carry them out. I'm very concerned about this, because the Administration confirmed that they did not engage with the private sector. So in your experience, are the cuts proposed in the budget research areas—is the private sector willing to simply start funding if the federal government cuts these?

Dr. SCHWARTZ. I shouldn't be speaking for the private sector. I gave one example earlier of when Ames Laboratory developed a new material, industry says okay, we got it. They didn't get it, and about 20 years later, we are reinvestigating. We are pursuing that path again. I am sure there are cases where private sector can pick some of it up. I don't think that that's going to be sufficient.

Ms. BONAMICI. I see that my time is expired, but I would like to follow up on that later.

Thank you, Mr. Chairman. I yield back.

Chairman WEBER. I thank the gentlelady for yielding back. Mr. Webster, you are up for five minutes.

Mr. WEBSTER. Thank you, Mr. Chairman. I would like to focus in on one thing, and that is a couple years ago there was sort of the storm of the century in the Northeast, and there was about $50 billion it cost the federal government to pay for the damages that were done there. Also back a few years ago—I have relatives in Chicago and in Oakridge, and I've toured both the national laboratories there. It seems like maybe one, maybe both were working on some fiber for composite material that would be way less expensive than what it is at that time, and that was—I was interested mainly in the construction industry because of resilient construction. I've been trying for a few years here—I did finally get resilient construction defined, so now we have it defined, and yet I could see the real potential with composite materials and construction areas, not only from a light weight, but also a durability so that when we have these storms, you know, our loss may have been in the hundreds of millions, but not $50 billion.

Could someone talk about—maybe Dr. Tirrell—of what's going on at the national laboratories in that research area?

Dr. TIRRELL. Thanks for that question. There's—that's one of many kinds of efforts in composite materials, some of which are based on additive manufacturing, some of which are based on new polymerization methods. Many of these things have organic plastic components to them. That's where the light weight comes from.

Mr. WEBSTER. Would that also—can I ask——

Dr. TIRRELL. Sure.

Mr. WEBSTER. Would that facilitate using these 3D printers——

Dr. TIRRELL. Yeah.

Mr. WEBSTER. Yeah.

Dr. TIRRELL. Yeah, that's what I was getting at, and I did—I wanted to say something earlier, too. I think there's huge frontiers on 3D printing. As 3D printing developed, it really wasn't 3D printing in a way. It was 2D printing over and over again. But now by the application of other kinds of fields of light and so on—I'm a polymer scientist myself, so I'm thinking more about the organic materials than the metals, but one can make very spectacularly different 3D shapes than could be made in the early days of 3D printing of polymers. There's startup companies in this area—but anyway, at Argonne, which is what I'm representing today, we're trying to open up a field that we call manufacturing science.

Mr. WEBSTER. By the way, Dr. Don Hillebrand gave me the tour.

Dr. TIRRELL. Good. Well he's the director of our energy system division.

Manufacturing science refers to the new science questions that come up. When you try to take something from the laboratory into larger scale production, you're doing it bigger, faster, cheaper, and the materials just don't behave the same way at that scale and at those time scales as they did in the lab. So Argonne is trying to be a leader in, as I said, what we're calling manufacturing science, which is new basic science applied to a manufacturing scenario.

Mr. WEBSTER. Are you familiar with the term resilient construction?

Dr. TIRRELL. Yes.

Mr. WEBSTER. The whole idea is that you can use the building the next day——

Dr. TIRRELL. Right, yeah.

Mr. WEBSTER. —once the wind comes or whatever comes.

Dr. TIRRELL. Yeah, resilience in general is a big focus at Argonne which extends beyond material science, but we're on materials here today, so——

Mr. WEBSTER. Well in other—along those same lines in science, there is—matter of fact, it seems like there's a couple universities offering corrosion engineering as a graduate degree, and it just seems like that—the construction, especially in maybe the realm of steel or other things where there's so much corrosion that there would be some usefulness in that.

Dr. TIRRELL. Yeah, absolutely. I mean, that's a huge economic drain. I mean, so far we've lived with it, but the point is if you could stop that or make materials last longer—and there are various centers of excellence. It's not a particular focus at Argonne.

Mr. WEBSTER. Great.

Dr. Higgs, too, I'd like to say to you come back to Georgia Tech. I just did the commencement exercise there here a few weeks ago, but you were a great contributor at that time. It's been a while.

But anyway——

Dr. HIGGS. Thank you.

Mr. WEBSTER. I—when I graduated as an engineer, my mom gave me a card that said four years ago, I couldn't even spell engineer. Then you open it up, on the inside it said now I are one, so——

Dr. HIGGS. Right.

Mr. WEBSTER. —I still are one, even though I've got a different profession now.

I yield back.

Chairman WEBER. Did she ask for any repayment of the money back?

Mr. WEBSTER. She should have.

Chairman WEBER. I understand. Our parents give us a lot, don't they?

Ms. Esty, you're now recognized for five minutes.

Ms. ESTY. Thank you, Mr. Chairman, and I want to encourage my friend, Daniel Webster, to join the Corrosion Prevention Caucus with me and Pete Olson, and the Resiliency Caucus, because we are very interested in these issues, and again, I think this is an area where basic research can save money, save lives, and would encourage that to be part of sort of our national initiative, and particularly with a move to pull us out of the Climate Accords. Climate is going to do what it's going to do. We need to be prepared, so I would encourage all my colleagues to do that.

I had a couple things I wanted to quickly go through in the limited time I have. First was just give an example that illustrates what many of my colleagues have talked about. I represent Connecticut. U–Conn has the Materials Genome Initiative funded through NSF. They're deeply worried. They came to meet with me a couple of weeks ago, and are very concerned about what these proposed cuts would do to their program, and many of those issues you've discussed about not only losing those particular projects, but in so doing, lose the talent pool, lose the grad students, lose the entire lab. And so I just think we really need to understand the implications. It's not a one-year cut. We actually risk losing them to other countries. We risk American competitiveness. So that's one. I just want to lend my voice to others.

The two other topics I want to quickly touch on, one is on ARPA–E, and the other is on STEM diversity and diverse workforce, which many of us are pretty passionate about.

Dr. Tirrell, I know that you've—the Argonne lab has done a lot of work on ARPA–E. If we're going to look at advanced materials and energy efficiency, it's incredibly important. You've done a lot of important work. We're looking at, you know, dramatic basically elimination of that. Could you talk a little bit about whether you think the private sector can fill in that gap, you know, the difference between who does basic research and who doesn't do basic research? I appreciate the mention, Dr. Higgs, of SBIR and that translation from basic research into commercial exploitation, but the basic research still has to be done. Dr. Tirrell, if you could talk a little bit about that.

Dr. TIRRELL. Well it does turn out that I am part of an ARPA–E project based at Argonne that has to do with how to improve the both acoustic and thermal insulation of windows with polymer coatings, and as I mentioned, I'm a polymer scientist. And so, you know, with a very well-defined need specified, we'd like to have this much insulation for sound and this much insulation for heat, and by the way, you can't make the windows foggy or anything like that. We're trying to design some polymers that will do that. So it's a good example of use-inspired basic research.

I also pointed earlier on to the basic energy science basic research needs workshops that in some ways frame things like that. They look at what an area of technology needs, and then talks about where we're missing out in basic research.

On the EERE or the Energy Efficiency and Renewable Energy, I think within that, there are great ways of advancing U.S. energy competitiveness. There's the Advanced Manufacturing Office, which relates to some of the things I was saying to Representative Webster about manufacturing science. So you know, I think these are valuable programs. I'll just leave it at that. They do things in a special way and produce good results.

Ms. ESTY. Thank you very much.

Dr. Locascio, I know you've recently blogged about diversity and science in your son's pride, and you being a scientist, and I was just with my big data son early this morning and thought about the importance of modeling that. And Dr. Higgs, you're noted for your efforts as well.

Quickly, for both of you, what can we do? What can the U.S. Congress do that would help ensure we are actually opening up that pipeline for each and every young person in this country to understand these are exciting fields? And we need their talent. We need their life experience. We need their input and their energy. Thanks.

Dr. LOCASCIO. Thank you for the opportunity to speak about this. I'm so passionate about it as well, so I appreciate that.

Yes, so there are several things that you talked about. First, getting people into the workforce is very difficult, and as you said, getting females or attracting females into the STEM research fields is very difficult. So given the fact that there could be changes in the way that we're recruiting and attracting people, at this particular time and in the budget, I think it makes it even more difficult. But the second part is retaining them, and then the third part is elevating them to a stature of leadership.

And so that's something that I have really thought a lot about. How do we make sure no matter what you look like or where you come from, what your cultural background is, we need you at the table in order to get the best people and the best ideas out there and supported for the sake of science in the United States. And so mentoring, guiding people, trying to make sure that we have adequate salaries to recruit them and retain them, they're all important facets of the equation. But then just making sure that we elevate them and promote them fairly, equally, and then showcase their talent in front of people so that they can be seen, I think is critical.

Dr. HIGGS. Very good question.

So I will definitely say that we like to produce a diverse number of scholars. A lot of you all have met goals because you've seen people that look like you, and it's the same dynamic that goes on with young people. I myself graduated from a historically black college and university. I saw people that looked like me had Ph.D.s and so I wanted to do that. I see my friend over here, Chris Jones, just got his Ph.D. from MIT. He's a graduate of Morehouse College as well. He saw people that looked like him, and he wanted to go and be an astronaut and do other things, like Mr. Webster become a politician and engineer as well. So it's a very important part of producing the nation's next generation of scientists and engineers. Thank you.

Ms. ESTY. Thank you very much.

Chairman WEBER. The Chair now recognizes Mr. Hultgren for five minutes.

Mr. HULTGREN. Thank you, Chairman. Thank you all so much for being here. This is really important, something we're passionate about, I'm passionate about, and research and development is so core, and especially that basic scientific research is something we've got to make sure funding is continued to remain, something the private sector can't do. It's something we're going to continue to fight with the current Administration and also fought the past Administration oftentimes where they were pushing certain types of projects and away from basic research. And so I want you to know there's strong voices on both sides of the aisle that have— continue that commitment and will continue to fight.

Also, I share my Illinois colleagues to thank Argonne. Thank you, Dr. Tirrell and the great work that Argonne is doing. We're so proud of you, so proud of what's happening at Argonne. But also at a time when there's not a lot to brag about in Illinois, we can brag about our research and so proud of Argonne and Fermi. You look at the data, the Elsevier and the Illinois Science and Technology Coalition. Rankings recently put Illinois ranking at 94th percentile in publication impact for material science fields, 86th percentile in publication volume. That's very impressive and something we absolutely are proud of. And I think a large recent we got that big impact is because the national labs accessibility certainly to students, but also as user facilities they are crown jewels in our research ecosystem. And that gives access to researchers throughout the country to high-end tools which no one university or business could ever maintain or have access to. So thank you. Keep up the great work. We're here to support you.

These user facilities are also proposed in a well thought out manner where the research community must set goals through the advisor committee process, and base these facilities on long-term needs. The 2016 BSAC report called the advanced photon source upgrade "absolutely central" to contribute to world leading science and ready to initiate construction.

Dr. Tirrell, I wonder if you could explain to the Committee why this facility upgrade is absolutely central to contribute to world leading science. Also, could you describe who the users are at such a facility? Where will this research be done, if not here in the United States?

Dr. TIRRELL. Thank you very much.

Yes, there's over 5,000 users every year of the advanced photon source. The upgrade is really necessary to keep it at the state of the art or push the state of the art. And by that, what we mean is intensity and coherence of the x-ray beam, and the more intense and the more coherent, the better—the more like a really infinitely powerful microscope the x-ray source becomes. So it sort of changes its nature a bit from a scattering tool to an imaging tool.

Investments are being made in Europe and in Japan, and they're pushing the frontiers too, but the APS upgrade will land us in 2025 with the best hard x-ray source in the world, and that will keep not on the U.S. science community strong itself, but it will keep people from all over the world coming here because we are the best. That's very enriching.

Mr. HULTGREN. It is, and that's, I think, the point that we always have to continue to come back to, remind ourselves certainly the value of these 5,000-plus users, the access that they have, the multiple impact on our economy for new discoveries there. I've heard about some amazing things that are coming out that really could be game changers for the world as far as energy goes, but also economic impact. So it is really important.

The other point you bring up is this research likely is going to happen, if not here, somewhere else. A lot of other countries are aggressive. They're not where we are. They're not able to lead right now, but if we fail, they're willing to step in. But we're also recognizing for us to be a part of important, big, groundbreaking, earth shattering research, collaboration likely is going to have to be a part of that. Reaching out and bringing other countries is part of that. I wonder if you could just talk a little bit about that, looking for solutions to new problems like new materials for batteries, or solving other problems in material science, how collaboration works within our own country. So Fermi Lab working with Argonne and University of Chicago at the Institute of Molecular Engineering for the Chicago Quantum Exchange, talking a little bit about these hubs, but then also how that's a draw on the international stage as well.

Dr. TIRRELL. Yeah, thanks very much.

You know, back on the thing that you said about Elsevier, I was actually contacted by a writer from Nature magazine who wants to write a story about material science in Northern Illinois, which is something I have been hoping for——

Mr. HULTGREN. Fantastic.

Dr. TIRRELL. —for a while. The Chicago Quantum Exchange is an effort to merge our resources among the institutions in Northern Illinois and in the Chicago area to lead in the next phase of what might be called post—computing, and that's, again, you know, a very, very competitive situation.

I have in front of me two weeks ago Science magazine that touts the Chinese communication satellite that demonstrated quantum communication between a satellite and Earth. You know, the world—the United States, you know, just went into really overdrive when Sputnik was launched in the '50s. That was launched by a country that was our adversary, but not in any kind of economic shape to drive developments. China is a whole different story. They are.

Mr. HULTGREN. They are, sir, right, and I think that is something that will be continuing to be motivating for us as Members of Congress, but also I think this Administration, that we can lead. We need to lead. We should lead. We're in the right spot, but we got to make sure that we're following it up with the proper support there.

I could go on for another 20 minutes. Thank you all for being here. We're so proud of you. Dr. Higgs, just want to give a shout out that grateful for your research, your work. I would say you're certainly an inspiration to many, and I would say—you talk about people who look like us, but I would say to all of us, all of you are inspirations. I just want to thank you for your great work. It is so important for us to inspire that next generation that science and discovery is still important, and it can happen here in America. So thank you. Keep up the great work. Let us know how we can help.

I yield back.

Chairman WEBER. I thank the gentleman. The gentleman surfer from California is recognized for five minutes.

Mr. ROHRABACHER. Thank you for acknowledging my great achievement. All right.

Chairman WEBER. It's the one time he can wax eloquent. Mr.

ROHRABACHER. There you go. Oh, that's good. I like that.

All right. Okay, first of all, let us know we wouldn't be on this Committee if we didn't believe in basic research. I mean, that's Republican, Democrat, we all are on this Committee; however, we are also Members of the House that have to deal with budgets, and it's great idealism. I happen to believe in limited government, and I believe how we can make sure that government doesn't grow out of proportion is making sure that science develops alternatives so that we can solve vexing problems through science rather than through bureaucracy. So nothing—let me just note, nothing should say that we are not united in that, but let me just note that when you're dealing with budgets, my colleagues on the other side of the aisle lament that we got out of the Paris Treaty, which really cost us billions of dollars, billions. That was the purpose of it was to redistribute wealth from us to other countries that weren't quite so well off. Now whether we like that or not, the fact is that means those billions wouldn't be available for us for scientific research. And so when we're talking about this, let's keep that in perspective, that there are other things people are complaining about, trying to have to deal with budgets across the board, which we try to do, that you can't ask for billions more to be spent on the Paris Treaty and expect to have full funding for these projects.

Let me ask, how do we get more money in from—we conferred to this with the space program about two decades ago when I was very involved in this Committee on that, and I—we figured out we couldn't put more money in and balance the budget in terms of the space program, and I'm very proud that I worked on the Commercial Space Act and with that Space Act, we laid the foundation for billions of dollars of private sector involvement in space. And that was the new resource that we had coming in. And is there some way that, number one, we can get the private sector—for example, right now these studies that you do and the information that you come forward with, the new materials that you're talking about that play such a vital role in progress, companies actually utilize this to build products that help our lives. But they also make a big product—I mean, a big profit in making those products. Do we have now a situation where those companies that are profiting by using your direct research in some way are paying a payback to the federal government or to this—our science community?

Dr. TIRRELL. The short answer is yes, they are, but not as much as they might.

In some ways, universities and national labs have filled in the gap for what used to be much more vigorous and extensive industrial research labs in the chemical industry, in the electronics industry, in the computer industry and so on, so you know, I think companies do, obviously, what's in their interest. That's what they're supposed to do. But I think it would be in their interest to invest more in collaborations with universities and national laboratories.

Mr. ROHRABACHER. Yeah. When I was young, my dad took me to that laboratory there in Dearborn, Michigan, and it was Edison's lab up there and it was really very impressive for me to see that. We went to—also next door to where they were developing new things for the cars. That was private funding, and I think Edison's was privately funded as well, come to think of it. Should—is there—we need to make sure that we do not encourage our industry to continue to be subsidized like this. If there is a way that someone is using the research, should we not try to make further demands on people? If they're going to make a profit from what you're researching, shouldn't they be paying more then for the use of that, instead of having the taxpayers having this as a hidden subsidy?

Dr. TIRRELL. Well I think, you know, it's a complicated situation. I don't think—at least I couldn't tell you what the right formula would be there. I would just express an overall hope that there would be more collaboration.

Mr. ROHRABACHER. Well if we do it for free, we can't blame the companies for taking it free. And we have a patent system in our country. Isn't—couldn't we then—is there a way that we could expand the protection of the patent so that materials that are developed in the public sector are—or even in the private sector, but mainly what you're doing with public money, that that has to be repaid to the owner of the patent, which would be the government in that case?

Dr. TIRRELL. Well generally speaking, at universities or at national labs, the owner of the patent is the university or the national laboratory, and then licensing fees are paid. And Argonne gets millions of dollars a year in licensing fees. So that kind of thing is happening——

Mr. ROHRABACHER. Okay.

Dr. TIRRELL. —and you know, I think it is a matter of developing a good system and figuring out if the balance is right there.

Mr. ROHRABACHER. Well let's see if we can do that. That's an avenue—we shouldn't just look at scientific basic research as simply it's going to be part of the federal bureaucratic programs that we—let's see if we can make things more efficient by making sure that the people in the private sector who profit from what you're doing are maybe paying a little higher share, but also, that will encourage them to be doing research as well.

So with that, thank you very much for all the good work you're doing. I certainly wish you success in coming up with a material that's going to make us cool in the summer and warm in the winter. That's great. Thank you very much.

Chairman WEBER. I thank the gentleman for yielding back.

I want to thank the witnesses for their valuable testimony and the Members for their questions today. The record will remain open for two weeks for additional comments and written questions from the Members.

I do want to end by saying that this Committee and the full Science Committee obviously is committed to research. Chairman Smith has been a staunch advocate of it, both sides of the aisle. And so we look at this budget and we say that is simply a submitted budget, but I'm going to encourage and I think we're going to continue to be able to help with research as much as absolutely possible. We are holding—trying to do a lot of things, spinning a lot of plates. If you all could quickly come up with a material to make those plates lighter, you know, it would make our job easier.

So I want to say thank you for being here today again. You all have—we could have gone on for a long time. This is very, very interesting. We appreciate what you guys do.

This hearing is adjourned.

[Whereupon, at 12:04 p.m., the Subcommittees were adjourned.]

Appendix I

ADDITIONAL MATERIAL FOR THE RECORD

STATEMENT SUBMITTED BY FULL COMMITTEE RANKING MEMBER
EDDIE BERNICE JOHNSON

OPENING STATEMENT
Ranking Member Eddie Bernice Johnson (D-TX)

House Committee on Science, Space, and Technology
Subcommittee on Energy
Subcommittee on Research and Technology
"Materials Science: Building the Future"
June 28, 2017

Thank you Chairman Weber and Chairwoman Comstock.

I would like to extend a special welcome to our witnesses and thank them for being here today. This is a subject I find truly fascinating, and I look forward to this opportunity to learn more about the value of federal support for materials science research. I also look forward to hearing your suggestions for how we in Congress can continue to enable progress in this vital field. Materials science is truly interdisciplinary, as it encompasses various aspects of physics, chemistry, engineering, and even biology in some cases.

Given the scientific challenges in advancing this field as well as its widespread potential applications, materials research is frequently and necessarily supported by the combined efforts of government, industry, and academia. To this end, federal agencies such as the Department of Energy's Office of Science, the National Science Foundation, ARPA-E, the National Institute of Standards and Technology, and DOE's Advanced Manufacturing Office play a key role by providing the critical funding needed by scientists and engineers in the field.

Every day we see the tangible results of this federal support. Only this month, researchers at the University of Michigan funded by NSF published a report on the development of a new semiconductor alloy that could revolutionize the solar power industry by lowering the costs of solar power dramatically. This new alloy is able to capture the sun's near-infrared light and is on track to achieve nearly 50% efficiency. Compare that to the less than 30% efficiency we see in traditional solar panels and you will understand why it has been called "magic" by some on the research team.

Simultaneously, Oak Ridge National Laboratory is working with the University of Wisconsin to develop a set of new heat exchangers which can be 3D printed to suit nearly any engineering design. The novel combination of copper fibers and nylon will have profound implications for power plant efficiency. I am certain I speak for all of my colleagues when I say that these scientists truly are a national treasure. It is largely through their efforts that the United States is able to remain a world leader in materials-related technology, and I applaud their dedication.

Finally, I must voice once again my deep dismay at President Trump's proposed budget cuts to our entire research and development enterprise. Given the examples I just provided, I find it ironic that we are even considering slashing these programs during what the White House is calling "Energy Week." As much as we might hope otherwise, a broad range of respected non-partisan institutions and experts have concluded that industry alone cannot and will not support of the majority of the research needed to pursue these long-term, high-risk innovations. Materials science is a field that employs nearly 8,000 Americans, and it would be a serious mistake to jeopardize these research positions and their potential for achieving new breakthroughs. I will continue to strongly support these activities and I encourage my colleagues to do the same.

Thank you, and I yield back.

LETTER SUBMITTED BY REPRESENTATIVE BILL FOSTER

Congress of the United States
Washington, DC 20515

June 9, 2017

Rick Perry
Secretary of Energy
U.S. Department of Energy
1000 Independence Ave, SW
Washington, DC 20585

Francis Collins, M.D., Ph.D.
Director
National Institutes of Health
9000 Rockville Pike
Bethesda, MD 20892

Scott Pruitt
Administrator
Environmental Protection Agency
1200 Pennsylvania Ave, NW
Washington, DC 20460

Dr. France Córdova
Director
National Science Foundation
4201 Wilson Blvd
Arlington, VA 22230

Dr. Kent Rochford
Acting Director
National Institute of Standards and Technology
Department of Commerce
100 Bureau Drive
Gaithersburg, MD 20899

Benjamin Friedman
Acting Administrator
National Oceanic and Atmospheric Administration
U.S. Department of Commerce
1401 Constitution Ave, NW
Washington, DC 20230

Robert M. Lightfoot, Jr.
Acting Administrator
National Aeronautics and Space Administration
300 E Street, SW
Washington, DC 20546

Dear Secretary Perry, Dr. Collins, Administrator Pruitt, Dr. Córdova, Dr. Rochford, Acting Administrator Friedman, and Acting Administrator Lightfoot:

We write to you today to ask for information on the number of jobs that would be eliminated at our nation's universities and National Laboratories if the President's Budget Request (PBR) for Fiscal Year 2018 were to be enacted.

Our universities and National Laboratory system support a world-class workforce of research scientists, engineers, and support personnel who work as teams on long-term solutions to address some of the nation's greatest challenges. Moreover, our universities and National Labs play a unique and critical role in educating the next generation of American scientists and researchers, including thousands of graduate students and postdoctoral researchers at hundreds of U.S. institutions who depend upon federal funding and facilities for their research and training.

We recognize that we must set priorities and make smart, strategic decisions about federal funding. But we also have serious concerns about the priorities this Administration appears to be setting. From economic stagnation and dependence on foreign energy to curing diseases and addressing threats to our national security, scientific research is the foundation for the innovative solutions that will enable us to overcome many of our greatest challenges. Instead, the funding levels requested in the FY18 PBR would cause permanent damage to our research infrastructure and workforce and furthermore force our National Labs to lay off critical scientific staff. This is no way to keep America great or maintain our position as a leader in science and innovation.

We cannot properly evaluate the full impact of the FY18 PBR until we know how many university researchers and National Laboratory employees would lose their federal funding or their jobs if these funding levels were enacted. Please provide your projections on how many researchers would be supported in the FY18 PBR compared to the current level as of the spring 2017. Also provide your estimates on how many National Laboratory employees would lose their jobs if these funding levels were enacted. We appreciate your detailed response no later than June 15, 2017.

Sincerely,

Bill Foster
Member of Congress

Kyrsten Sinema
Member of Congress

Grace F. Napolitano
Member of Congress

Derek Kilmer
Member of Congress

Raúl M. Grijalva
Member of Congress

Anthony Brown
Member of Congress

Ron Kind
Member of Congress

Ed Perlmutter
Member of Congress

Donald S. Beyer Jr.
Member of Congress

Debbie Wasserman Schultz
Member of Congress

Jackie Speier
Member of Congress

Paul Tonko
Member of Congress

Suzan DelBene
Member of Congress

Luis V. Gutiérrez
Member of Congress

Steve Cohen
Member of Congress

Jan Schakowsky
Member of Congress

Barbara Lee
Member of Congress

Jerry McNerney
Member of Congress

Betty McCollum
Member of Congress

Gwen Moore
Member of Congress

Gerald E. Connolly
Member of Congress

Gregory W. Meeks
Member of Congress

Mark DeSaulnier
Member of Congress

Eric Swalwell
Member of Congress

John Yarmuth
Member of Congress

Joyce Beatty
Member of Congress

Thomas R. Suozzi
Member of Congress

Mike Doyle
Member of Congress

Anna G. Eshoo
Member of Congress

Jacky Rosen
Member of Congress

Colleen Hanabusa
Member of Congress

Denny Heck
Member of Congress

Jamie Raskin
Member of Congress

James P. McGovern
Member of Congress

Dwight Evans
Member of Congress

Earl Blumenauer
Member of Congress

Zoe Lofgren
Member of Congress

Kathleen M. Rice
Member of Congress

Jared Polis
Member of Congress

Marcy Kaptur
Member of Congress

Sheila Jackson Lee
Member of Congress

Terri A. Sewell
Member of Congress

David N. Cicilline
Member of Congress

Danny K. Davis
Member of Congress

Mike Quigley
Member of Congress

Adam B. Schiff
Member of Congress

Pramila Jayapal
Member of Congress

David Scott
Member of Congress

Donald Norcross
Member of Congress

Darren Soto
Member of Congress

Robert C. "Bobby" Scott
Member of Congress

Matthew Cartwright
Member of Congress

Yvette D. Clarke
Member of Congress

Eleanor Holmes Norton
Member of Congress

John Garamendi
Member of Congress

A. Donald McEachin
Member of Congress

www.ingramcontent.com/pod-product-compliance
Lightning Source LLC
Chambersburg PA
CBHW082351220526
45470CB00008B/2712